SpringerBriefs in Applied Sciences and Technology

Thermal Engineering and Applied Science

Series Editor

Francis A. Kulacki, University of Minnesota, USA

For further volumes:
http://www.springer.com/series/10305

Patrick H. Oosthuizen
Abdulrahim Y. Kalendar

Natural Convective Heat Transfer from Narrow Plates

 Springer

Patrick H. Oosthuizen
Department of Mechanical
  and Materials Engineering
Queen's University
Kingston, ON
Canada

Abdulrahim Y. Kalendar
Mechanical Power and Refrigeration
  Department
College of Technological Studies
Shuweikh
Kuwait

ISSN 2193-2530   ISSN 2193-2549 (electronic)
ISBN 978-1-4614-5157-0   ISBN 978-1-4614-5158-7 (eBook)
DOI 10.1007/978-1-4614-5158-7
Springer New York Heidelberg Dordrecht London

Library of Congress Control Number: 2012945484

Printed on acid-free paper

Springer is part of Springer Science+Business Media (www.springer.com)

Preface

Natural convective heat transfer from wide flat vertical plates, i.e., from plates whose horizontal width is relatively large compared to their vertical height, has been extensively studied over many years. However, situations frequently arise in engineering practice that effectively involve natural convective heat transfer from plates that are relatively narrow, i.e., plates whose horizontal width is comparable to or smaller than their vertical height. In such cases the mean heat transfer rate can be much greater than that from wide plates of the same height, the increase often being said to be due to "edge" effects. However, it is only comparatively recently that relatively general equations have been derived that allow heat transfer rates in such narrow plate situations to be calculated. The heat transfer rate from a narrow plate can depend on the conditions existing at the sides of the plate. For example, it can depend on whether the plate is recessed within or protrudes from a surrounding adiabatic surface. And it can depend on the conditions existing at the plate surface, i.e., for example, on whether the plate is isothermal or whether there is a uniform rate of heat generation over the plate surface. These effects on natural convective heat transfer from narrow flat plates have also only relatively recently been extensively investigated. The magnitude of the edge effect is also dependent on whether the flow is in the laminar, transitional, or turbulent flow regions and studies of this have only recently become available. When the plate is not vertical, but rather inclined to the vertical, pressure changes normal to the plate surface arise. These pressure changes can alter the nature and the magnitude of the edge effects and these effects of plate inclination on the heat transfer rate from narrow plates have also only been recently studied to a significant extent. While most attention has been given to the heat transfer from single plates when there are two narrow rectangular flat plates of the same size separated vertically or horizontally but relatively close together, the flow interaction between these heated plates can also have a significant effect on the nature of the edge effect, i.e., on the heat transfer rates from the plates. Again, this effect has only recently been extensively studied. Thus, a number of both numerical and experimental results that allow the natural convective heat transfer rate from narrow plates in various situations to be calculated have recently become available. This book presents reviews of some of these results.

This book attempts, therefore, to provide information about the effect of plate aspect ratio (width to "height" ratio) on the natural convective heat transfer from flat plates in a variety of situations that can occur in practice. This book will hopefully provide background information for researchers in the field and assist practitioners in determining the heat transfer rates in various situations that they encounter in their work.

Kingston, Canada   Patrick H. Oosthuizen
Shuweikh, Kuwait   Abdulrahim Y. Kalendar

Contents

Symbols

A	Heat transfer area
C	Specific heat of material from which experimental model is made
g	Gravitational acceleration
G	Distance between two adjacent heated plates
Gap	Vertical distance between two adjacent heated plates
h	Height of heated plate
h_t	Total heat transfer coefficient
h_c	Convective heat transfer coefficient
h_{cd}	Conductive heat transfer coefficient
h_r	Radiation heat transfer coefficient
HGap	Dimensionless distance between two adjacent plates
I	Current passing through the heater
V	Voltage drop across the heater
K	Thermal resistance of Plexiglas base
k	Thermal conductivity of fluid
Nu	Mean Nusselt number based on h and on $(T_H - T_F)$
Nu_L	Local Nusselt number based on h and on $(T_H - T_F)$
Nu_y	Local Nusselt number based on y and on $(T_H - T_F)$
Nu_0	Mean Nusselt number when edge effects are negligible
Nu_{top}	Mean Nusselt number from the top plate
Nu_{bottom}	Mean Nusselt number from the bottom plate
Nu_{total}	Mean Nusselt number from both top and bottom plates
Nu_{mviemp}	Mean Nusselt number given by correlation equation for vertical and inclined plate facing up and facing down
Nu_{memp}	Mean Nusselt number given by correlation equation
Nu_{mc}	Mean Nusselt number for entire cylinder
Nu_{mcemp}	Mean Nusselt number given by correlation equation
P	Dimensionless pressure
p	Pressure
p_F	Pressure in fluid

Pr Prandtl number

$\bar{q}'$ Mean heat transfer rate per unit area

q'_y Local heat transfer rate per unit area

q'_w Heat flux at plate surface

Q_{conv} Total rate of heat loss by convection

Q_{cond} Total rate of heat loss by conduction

Q_{rad} Total rate of heat loss by radiation

Ra Rayleigh number based on h and on $(T_H - T_F)$

Ra_y Rayleigh number based on y and on $(T_H - T_F)$

Ra^* Heat Flux Rayleigh number based on h and on q'_w

Ra_y^* Heat Flux Rayleigh number based on y and on q'_w

t Dimensionless height of plate "above" adiabatic surrounding surface t'/h

t' Height of plate "above" adiabatic surrounding surface

T Temperature

T_F Temperature of fluid

T_H Temperature of plate

$\bar{T}_H$ Mean plate temperature

U_X Dimensionless velocity component in X direction

u_x Velocity component in x direction

u_r Reference velocity

U_Y Dimensionless velocity component in Y direction

u_y Velocity component in y direction

U_Z Dimensionless velocity component in Z direction

u_z Velocity component in z direction

$VGap$ Dimensionless distance between two vertically separated plates Gap/h

w Width of the plate

W Dimensionless width of plate, w/h

X Dimensionless horizontal coordinate normal to plate

x Horizontal coordinate normal to plate

Y Dimensionless vertical coordinate

y Vertical coordinate

Z Dimensionless horizontal coordinate in plane of plate

z Horizontal coordinate in plane of plate

Greek Symbols

α Thermal diffusivity
β Bulk expansion coefficient
φ Angle of inclination from the vertical
ν Kinematics viscosity
θ Dimensionless temperature
σ Stefan–Boltzman constant
ε Emissivity of the model

Chapter 1
Introduction

Keywords Natural convection · Narrow plates · Wide plates · Inclined plates · Adjacent plates · Literature review

1.1 Introduction

Two-dimensional natural convective heat transfer from a wide vertical isothermal plate has been extensively studied and equations for the prediction of the heat transfer rates that occur in this situation have been developed. However, when the width of the plate is relatively small compared to its height, i.e., when the plate is narrow (see Fig. 1.1), the heat transfer rate can be considerably greater than that predicted by these two-dimensional flow results. The increase in the heat transfer rate from a narrow plate relative to that from a wide plate under the same conditions results from the fact that fluid flow is induced inwards near the edges of the plate and the flow near the edges of the plate is, thus, three-dimensional (see Fig. 1.2). When the plate is wide these edge effects are negligible. However, when the plate is narrow the edge effects on the heat transfer rate can be quite significant. The increase in the heat transfer rate with a narrow plate is often said to be due to "three-dimensional edge effects" or simply "edge effects".

Situations that can be approximately modeled as narrow vertical plates occur in a number of practical situations, so there exists a need to be able to predict natural convective heat transfer rates from such narrow plates. While there have in the past been some limited studies of the natural convective heat transfer rate from narrow plates (see discussion below), the results obtained in these studies have been for a relatively narrow range of the governing parameters and there has existed until relatively recently a need for a broader range of results for this situation that can be used as the basis for predicting heat transfer rates from narrow plates in practical situations. In the recent numerical studies reviewed in this book, results for the

P. H. Oosthuizen and A. Y. Kalendar, *Natural Convective Heat Transfer from Narrow Plates*, SpringerBriefs in Thermal Engineering and Applied Science, DOI: 10.1007/978-1-4614-5158-7_1, © The Author(s) 2013

Fig. 1.1 Wide and narrow
vertical plates

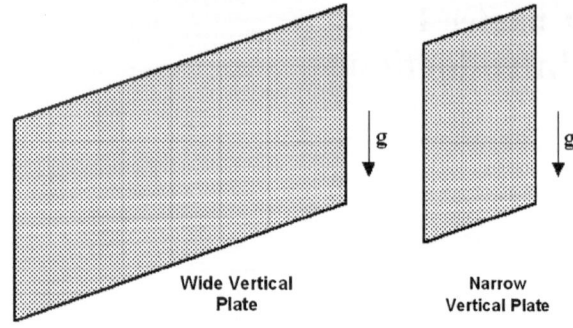

Fig. 1.2 Flow near edge of
plate

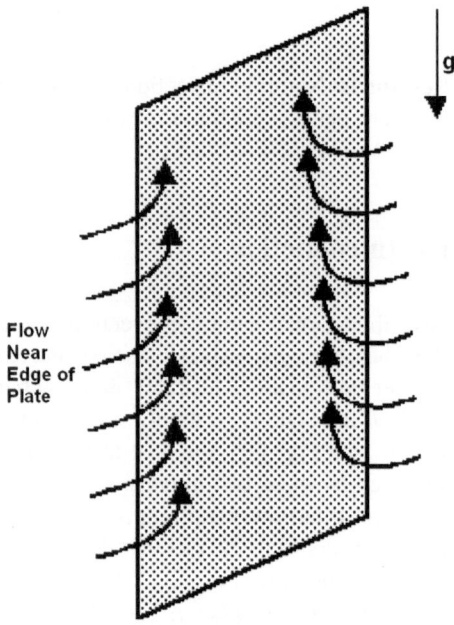

natural convective heat transfer rates from narrow vertical and inclined isothermal
plates for a relatively wide range of Rayleigh numbers and dimensionless plate
widths have been numerically determined in various situations. In all cases considered, attention has been restricted to results for a Prandtl number of 0.7, this
being approximately the value for air and thus the value existing in the applications that originally motivated these studies. The basic flow situation that has been
considered in most of the studies considered in the present book is shown in
Fig. 1.3. The narrow heated plate is, as shown in this figure, surrounded by an
adiabatic surface that is in the same plane as the heated plate. The width of the
plate, w, is assumed to be less than the vertical height of the plate, h.

Fig. 1.3 Situation
considered in most of the
work described in the present
book

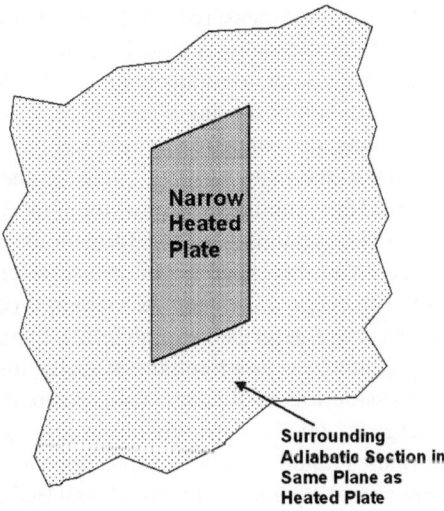

Narrow
Heated
Plate

Surrounding
Adiabatic Section in
Same Plane as
Heated Plate

The purpose of the present chapter is to give a brief review of studies of natural convective heat transfer from wide plates and to briefly discuss work on heat transfer from narrow plates.

1.2 Wide Vertical Flat Plates

Most existing studies on natural convective heat transfer from flat plates have considered the situation where the plate is wide compared to the height. This means that the flow can be assumed to be two-dimensional, i.e., edge effects can be ignored. Many authors have investigated natural convection heat transfer from wide vertical and inclined flat plates in both the laminar and turbulent flow regions over the past 100 years. Some of the more important of these studies that are related to the present work are reviewed in this section.

In 1915, Nusselt mathematically set up the conservation equations for mass, momentum, and energy for the natural convection heat transfer case. He undertook the first investigation in which the heat transfer data were expressed in terms of dimensionless parameters, i.e., in terms of what are now known as the Nusselt, Nu, Grashof, Gr, and Prandtl, Pr, numbers. He showed that the relations for natural convective heat transfer whether local or average could be expressed in the following form:

$$Nu = f\left(Pr \times Gr\right) \tag{1.1}$$

Early experiments on natural convective heat transfer from vertical wide flat plates in air were undertaken by Schmidt and Beckman [1]. In their studies, the temperatures in the boundary layer were measured by thermocouples and the velocities were measured using a quartz thread anemometer as discussed by Squire [2]. The results were obtained for flat plates having heights of 12.5 cm and 50 cm which were kept at a temperature of 65 °C in air at 15 °C. They also obtained an analytical solution using a simplified form of the governing equations obtained by neglecting certain terms, particularly by assuming that the velocity component normal to the plate was negligible. At the same time, Pohlhausen (see Squire [2]) reduced the simplified partial differential equations to ordinary differential equations by applying similarity transformations. His predicted temperature distributions agreed with the experimental measurements but there were significant differences between the predicted and the measured velocity distributions. A general method of solution was described by Squire [2]. In his work, a suitable third-order polynomial expression was used for the velocity distribution and second-order polynomial expression was used for the temperature distribution in the integral momentum and energy equations. As discussed in [2], Saunders and Schuh obtained solutions for various Prandtl numbers using approximate methods. The simplified ordinary differential equations of Schmidt and Beckman [1] were solved for different Prandtl numbers by Ostrach [3] using an early computer system. He seems to have been the first to present the exact numerical similarity solutions for free convective boundary layer flows over a wide isothermal vertical flat plate. The authors whose work was discussed above obtained expressions for local and average Nusselt numbers in laminar natural convective heat transfer from a vertical flat plate in terms of local Grashof number and Prandtl number. These expressions have the form:

$$Nu_y = K_1 \left(Pr \times Gr_y \right)^{0.25} \tag{1.2}$$

$$Nu = K_2 \left(Pr \times Gr \right)^{0.25} \tag{1.3}$$

where K_1 and K_2 are constants. Different values of the two K constants were derived in the studies mentioned above. More detailed and general expressions were obtained by Eckert [4].

An exact solution of the laminar boundary layer equations for natural convection over a wide vertical plate having uniform surface heat flux was obtained by Sparrow and Gregg [5]. The results of their experiments for this situation were found to be in good agreement with the theoretical results.

A widely used empirical correlation equation based on experimental results to estimate the average Nusselt number for an isothermal wide flat plate for the entire Rayleigh number range of laminar, transition, turbulent flow regions, and for different Prandtl number was constructed by Churchill [6] and Churchill and Chu [7]. A comprehensive review of these studies is presented in many heat transfer books such as Holman and White [8], Burmeister [9], Bejan [10], Kakac and Yener [11], Oosthuizen and Naylor [12], Lindon and Thomas [13], William et al. [14], Ioan and Derek [15], Oleg and Pavel [16], Mickle and Marient [17], Incropera et al. [18], Jaluria [19], McAdams [20]. Most of these books also discuss commonly used methods based on the similarity solution or on the integral equation method which have been used for calculating natural convective heat transfer for conditions under which the boundary layer equations apply.

1.3 Inclined Wide Flat Plates

Natural convective heat transfer from wide flat plates oriented at various angles to the vertical has been investigated by many authors. Rich [21] appears to have undertaken one of the first studies of this situation. He obtained local heat transfer rates in the laminar flow region for an inclined plate 101.6 mm in width × 406.4 mm in height. Rich determined the temperature field using a Mach–Zehender interferometer for local Grashof numbers in the range of 10^6–10^9. The inclination angle of the heated plate was between $0°$ and $40°$ to the vertical. His results indicated that the heat transfer coefficient for an inclined heated plate could be predicted using equations for a vertical plate if the gravitational term in the Grashof number was taken as component parallel to the inclined surface. Michiyoshi [22] applied the approximate integral equation method to natural convective flow over an inclined, isothermal, infinitely wide (which meant that the flow was two-dimensional), thin flat plate with a Prandtl number of fluid close to unity. Michiyoshi found that the average heat transfer coefficient of the lower surface is larger than that of the upper surface, and that the average heat transfer coefficient increases as the plate approaches the vertical position. In this analysis, he neglected the effect of the longitudinal pressure gradient in the boundary layer induced by the buoyancy forces and his results therefore, cannot be expected to be accurate for cases where the plate is near the horizontal. Kierkus [23] used a perturbation analysis of the boundary layer equations to obtain solutions for the velocity and temperature fields for laminar free convection about an inclined wide isothermal plate. Using a small parameter that involves the surface inclination and the Grashof number as the perturbation parameter and using the similarity solution for an isothermal vertical plate as the zeroth-order approximation, he determined a first-order approximate solution for angles up to $45°$ for both positive and negative plate inclination angles. The velocity, temperature, and local heat transfer results and Kierkus [23] support Rich's [21] conclusion. Vliet [24] seems to be the first who presented experimental local heat transfer results for natural convection over constant heat flux inclined surfaces using water and air as the test fluids. The data was for Rayleigh numbers from 10^7 to 10^{16} and inclination angles from vertical to $30°$ from the horizontal and included results for laminar, transitional, and turbulent flow regions. In this study, false walls were attached to each side of the heated plate in an effort to reduce edge effects. In the laminar flow region, the data was found to correlate well with vertical plate equations when the gravitational component parallel to the surface was used and this result supports Rich's [21] conclusion. Transition from laminar to turbulent flow was found to be strongly affected by the inclination angle, the transition value of $Gr_x \times Pr$ decreasing from

10^{13} for vertical surface to approximately 10^8 for a surface inclined at an angle of 30° to the horizontal. An experimental study of natural convection flow over inclined isothermal flat plates was also conducted by Hassan and Mohamed [25]. Their experiments were performed in air and their results cover a wide range of plate inclination angles from a horizontal heated plate facing upward to a horizontal heated plate facing downward. For conditions where it was possible to compare their results with existing results, good agreement was obtained. Laminar natural convection flow over an inclined isothermal heated surface was also studied by Lee and Lock [26]. They numerically solved the appropriate boundary layer equations for the flow of air and they presented graphically the effects of inclination angles on heat transfer, temperature, pressure, and velocity profiles.

Fuji and Imura [27] undertook an experimental study concerning natural convective heat transfer from heated plates of finite width (30 cm height × 15 cm width and 5 cm height × 10 cm width) immersed in water at arbitrary inclination angles. Their heated surface was neither isothermal nor was there a uniform heat flux at the surface. The main flow in the boundary layer was restricted to two-dimensionality by installing plates along both side edges of the heated plate and their study was for Rayleigh numbers between 10^5 and 10^{11}. They found Nusselt numbers for the smaller plate were somewhat larger than those for the plate with the larger height. For the inclined heated plate facing upwards, the wall temperature was found to be more uniform than that for the inclined plate facing downwards and the larger the angle of inclination became the smaller the transition Rayleigh number was found to be. The cause of the observed variation of heat transfer coefficient with inclination angle was ascribed to a change in flow pattern in the boundary layer. They also correlated their results in the laminar and turbulent flow regions. In the laminar flow region, the expression for a vertical plate was found to be applicable to the inclined plate if the gravitational term in the Gr number was altered to equal the component parallel to the inclined surface.

Vliet [24] performed an experimental study of the local heat transfer from an inclined upward facing heated plate with a uniform surface heat flux to the water. The main flow in the boundary layer was restricted to be two-dimensional and the maximum angle of inclination was 60° to the vertical. Vliet's results agreed with those of Rich [21] and correlate well in the laminar region with the vertical plate theory when the gravity component along the surface was used. In the turbulent flow case, the correlated results were independent of the inclination angle. Vliet and Ross [28] studied the local heat transfer for turbulent natural convection on vertical and inclined upward and downward facing surfaces. Their test surface was wide with a uniform surface heat flux and the results were obtained in air. The results showed the location of transition to be a function of the plate angle. Their laminar flow results were correlated using the gravity component along the surface and their results agree with those of Rich [21]. In the turbulent flow case, the correlation is independent of angle for the upward facing plate case whereas for the stable case (plate facing down) their data correlated best when the gravity is modified by $\cos^2\theta$.

Black and Norris [29] used a differential interferometer to provide flow visualization and measurements of the local heat transfer coefficient for free convection from an inclined isothermal plate. They found that the flow structure within the turbulent thermal boundary layer can be separated into a thermal sublayer and a core region that contains random fluctuations. The thermal sublayer was shown to contain "thermal waves" that traverse the heated surface and cause significant variations in local heat transfer coefficient and an overall increase in the heat transfer rate in the transition flow region but did not affect the results in the turbulent flow region. Their results for the local Nusselt number agreed with those obtained by other researchers.

Gryzagoridis and Klingenberg [30] experimentally studied the natural convective flow over an inclined isothermal plate in air using a plate having dimensions of 200 mm in width, 100 mm in height, and 8 mm in thickness. Their results show that for plate inclinations up to 60° from the vertical, the plate can be treated as vertical provided that the gravitational force component parallel to the surface is used. For angles between 60° and 75° a gradual change is noted, while for angles above 75°, the heat transfer was found to exhibit a definite change from that given by the vertical plate relation. It was also found that the upper and lower surface can be treated as similar for all angles below 60° from the vertical. Their results give the local Nusselt number at a chosen distance from the leading edge.

Laminar natural convection heat transfer from inclined plates with a constant heat flux was investigated by King and Reible [31]. Their heated surface was 0.1524 m wide and 0.3048 m high in water. Edge effects were avoided by applying a wall on each side edge to prevent the inflow from the sides. Their results agreed with the correlation predicted by a similarity solution for flat plate by applying $g\cos\theta$ as a gravitational component.

Wei et al. [32] conducted a numerical study to investigate the two-dimensional natural convective heat transfer from a uniformly heated thin plate with arbitrary inclination angle. The concurrent heat transfer from both sides of the heated plate was studied. They found that empirical expressions for the local and average Nusselt numbers were complicated and that the average Nusselt number cannot be correlated by a single equation for inclination angles less than 10° from horizontal. However, for inclination angles greater than 10° a correlation equation for the average Nusselt number could be obtained.

Turbulent natural convection on vertical and inclined upward and downward facing surfaces was investigated by Vliet and Ross [28]. Their heated plate had a constant surface heat flux and was 1.83 m wide and 7.32 m height. Results were obtained for inclination angles of from 30° to the vertical (upward facing) through the vertical to 80° to the vertical (downward facing), and the test were conducted in air with modified Grashof numbers up to 10^{15}. These results showed the location of transition to be a function of the plate angle and they correlated the local Nusselt number for both orientations using a single equation in which the gravity is taken as the component along the surface. In the turbulent flow region, they correlate their results for the local Nusselt number in a single equation where the gravity component is modified by $g\cos^2\theta$ for the case of downward facing plate while the correlation is independent of the angle of inclination for upward facing plate cases. Local and average heat transfer results for natural convection heat transfer from isothermal vertical and inclined plates facing upwards to air were experimentally investigated by Al-Arabi and Sakr [33]. Their experiments used a brass plate that was 1300 mm in height and 650 mm wide in air and considered inclination angles from 0° to 90°. Their results cover both the laminar and turbulent flow regions. From their results they obtained the critical Rayleigh number for the turbulent transition region and obtained local and average heat transfer correlations for both laminar and turbulent flows.

Other studies of natural convective flow over inclined plates are described by Jeschke and Beer [34], Kimura et al. [35], Kimura et al. [36], Komori et al. [37], Lloyd and Sparrow [38], Sparrow and Husar [39], and Sparrow et al. [40].

1.4 Narrow Flat Plates

Experimental studies of natural convective heat transfer from narrow isothermal vertical plates in the laminar flow region were undertaken by Oosthuizen [41]. His results showed that with a plate having a width that is comparable to the height, there is a considerable increase in the heat transfer rate relative to that which would occur if the flow was two-dimensional. Further experimental work was undertaken by the same author [42] that provided additional direct evidence of this plate width effect. Oosthuizen and Paul [43] predicted the magnitude of the edge effect for narrow vertical plates by numerically solving an approximate form of the boundary equations with additional terms for the transverse flow in the energy and momentum equations. Oosthuizen [44] studied edge effects on forced and free convective laminar boundary layer flow over a flat plate. The study was based on the use of the dimensionless boundary layer equations. The results were used to deduce the effects of the presence of the edge on the local and mean dimensionless heat transfer rates and to indicate the extent of edge region.

A solution to the three-dimensional Navier–Stokes equations for laminar natural convection over an isothermal narrow vertical flat plate was obtained by Noto and Matsumoto [45–47]. From their studies of the three-dimensional streamline patterns, of the local Nusselt number distributions, and of the wall vorticity distributions, they found that three-dimensional effects were quite evident in the natural convective field at low Grashof numbers. However, a limited range of the governing parameters were covered in their studies making it difficult to draw general conclusion from their results.

Oosthuizen and Paul [48, 49] describe numerical studies of natural convective flow over narrow vertical flat plates with a uniform surface temperature and with a uniform surface heat flux respectively, the heated plates being imbedded in a large plane adiabatic surface. The dimensionless plate width was shown to have a significant influence on the mean Nusselt number. This effect was shown to increase with decreasing dimensionless plate widths and with decreasing Rayleigh numbers.

An empirical equation for the mean heat transfer rate from narrow plates was derived from the numerical results. Studies of heat transfer from narrow plates that are inclined to the vertical are described by Kalendar and Oosthuizen [50] and [51].

Most of the studies mentioned above have been concerned with situations in which laminar flow exists. Heat transfer from narrow plates when turbulent flow exists has been numerically studied by Kalendar and Oosthuizen [52].

The conditions at the edges of a narrow plate can be expected to affect the magnitude of the edge effect on the heat transfer rate. Oosthuizen and Paul [53] studied natural convection heat transfer from a narrow vertical isothermal flat plate with different edge conditions. It was found that the edge conditions only affect the heat transfer rate when the Rayleigh number and dimensionless plate width are small. Other studies of the effect of edge conditions on natural convective heat transfer from narrow vertical plates are described by Oosthuizen and Paul [54–56].

As already mentioned, experimental studies of natural convective heat transfer from narrow vertical plates in the laminar flow region were undertaken by Oosthuizen [41, 42], his results showing that with a plate having a width that is comparable to the height, there is a considerable increase in the heat transfer rate relative to that which would occur if the flow was two-dimensional. Baker [57, 58] experimentally studied the effect of an electronic component's width to height ratio on free and forced convection from resistor chips with surface areas between 2 cm^2 and 0.01 cm^2. The heat transfer rate was obtained at the mid-point of the plate with the plate set vertically in air, liquid Freon-113, and silicon dielectric liquid. The heat transfer rates obtained were much larger than the values predicted by the two-dimensional theory. In fact, Baker's analysis indicated that for sources as small as IC chips the average convection coefficient might be more than an order of magnitude greater than that given by two-dimensional results under the same condition. It was suggested that these increases were due to side flow effects and it was found that the Nusselt number increased as the chip width decreased.

An experimental study of the free convection heat transfer rates from flush mounted and protruding microelectronic chips was undertaken by Park and Bergles [59]. In their study, the microelectronic chips were simulated by thin foil heaters. It was found that the Nusselt number increased as the chip width decreased. Even with the widest of the heaters used, the heat transfer coefficients were found to be higher than the values predicted by two-dimensional theory. However, the conditions at the vertical edges of the plates used in most of these past experimental studies were different from those assumed in the studies discussed in this book and the results obtained in the present study cannot, therefore, be quantitatively compared with those obtained in these earlier experimental studies. The range of parameters covered in these past experimental studies was also rather limited. For this reason, some newer experimental results for natural convective heat transfer from a narrow plate obtained by Kalendar and Oosthuizen [60] will also be discussed in this book.

1.5 Interaction Between Heated Plate Flows

Most existing work on natural convective heat transfer rates from plates has been focused on heat transfer from a single isolated plate. However, the interaction between the natural convective flows over narrow adjacent heated plates and its effect on the heat transfer rate from the plates is of considerable practical importance. The nature of the flow interaction between the flows over narrow plates can become quite complex when the plates are inclined at an angle to the vertical and this has not in the past been extensively studied. Here, the heat transfer that occurs when there is an interaction of the flows over adjacent plates is considered.

An experimental study of the free convection heat transfer rates from flush mounted microelectronic chips on a circuit board substrate and from protruding microelectronic chips that were protruding from the substrate by approximately 1 mm was undertaken by Park and Bergles [61]. In their study, the microelectronic chips were simulated by thin foil heaters with constant surface heat flux. In addition to measurements with a single heater the heat transfer from arrays of flush mounted heaters were measured with various distances between the heaters. The heat transfer coefficient for the top heaters were found to increase as the distance between the heaters increased, tending to a constant value at large distances.

There have been a number of other studies of natural convective heat transfer rates from various arrays of heated elements, typical of such studies being those described by Heindel [62], Tou et al. [63], Tou and Zhang [64], Baskaya et al. [65], Oosthuizen and Paul [66] and Kirby and Fleischer [67]. In these studies the effect of Rayleigh number, aspect ratio of the heaters and the Prandtl number on the mean heat transfer rate from the heaters was studied. Recent work on the interaction between the flows over adjacent narrow plates is described by Kalendar and Oosthuizen [68, 69] and Kalendar et al. [70].

1.6 The Present Book

The following topics involving natural convective heat transfer from narrow plates that are dealt with in the present book are: Natural Convective Heat Transfer from Vertical Narrow Flat Plates, Natural Convective Heat Transfer from Inclined Narrow Flat Plates, Effect of Plate Edge Conditions on Natural Convective Heat Transfer from Narrow Flat Plates, Experimental Results for Natural Convective Heat Transfer from Narrow Flat Plates, Natural Convective Heat Transfer from Narrow Flat Plates with Transitional and Turbulent Flow, and Natural Convective Heat Transfer from Adjacent Narrow Plates. The work described in the book is mainly based on recent studies by Kalendar and Oosthuizen [50–52, 60, 61, 69], Kalendar et al. [70], Oosthuizen, and Oosthuizen and Paul [48, 49, 53–56].

References

1. Schmidt E, Beckmans W (1930) Das temperature and geschwindigkeitsfeld von einer warmeabgebenden senkrechten platte bei naturlicher konvecktion. Tech Mech Thermodynam 1(341–349):391–406
2. Squire HB (1956) Modern developments in fluid mechanics: high speed flow, 1st edn. Oxford University Press, London
3. Ostrach S (1953) An analysis of laminar free-convection flow and heat transfer about a flat plate parallel to the direction of the generating body force. NACA report 1111
4. Eckert ER (1959) Heat and mass transfer, 2nd edn. McGraw-Hill, New York
5. Sparrow EM, Gregg JL (1955) Laminar free convection from vertical plate with uniform surface heat flux. Proceedings of ASME diamond jubilee semi-annual meet. Paper 55-SA-4
6. Churchill SW (1973) A correlation for laminar free convection from a vertical plate. Trans Am Soc Mech Eng Series C J Heat Transf 95(4):540–541
7. Churchill SW, Chu HHS (1975) Correlating equations for laminar and turbulent free convection from a vertical plate. Int J Heat Mass Transf 18(11):1323–1329
8. Holman JP, White PRS (1992) Heat transfer, 7th edn. McGraw-Hill, New York
9. Burmeister LC (1993) Convective heat transfer, 2nd edn. Wiley, New York
10. Bejan A (1995) Convection heat transfer, 2nd edn. Wiley, New York
11. Kakac S, Yener Y (1995) Convective heat transfer, 2nd edn. CRC Press LLC, Boca Raton
12. Oosthuizen PH, Naylor D (1999) Introduction to convective heat transfer analysis, 1st edn. McGraw-Hill, New York
13. Thomas LC (2000) Heat transfer, 2nd edn. Capstone Publishing Corporation, Tulsa
14. Kays W, Crawford M, Weigand B (2005) Convective heat and mass transfer, 4th edn. McGraw-Hill Higher Education, Boston
15. Pop I, Ingham D (2001) Convective heat transfer, 1st edn. Pergamon Press, New York
16. Oleg GM, Pavel PK (2005) Free-convective heat transfer, 1st edn. Springer, Berlin
17. Mickle F, Marient ST (2009) Convective heat transfer, 1st edn. Wiley and ISTE, London and Washington
18. Incropera F, Dewitt D, Bergman T, Lavine A (2006) Fundamentals of heat and mass transfer, 6th edn. Wiley, New York
19. Jaluria Y (1980) Natural convection: heat and mass transfer, 1st edn. Pergamon Press, New York
20. McAdams WH (1954) Heat transmission, 3rd edn. McGraw-Hill, New York
21. Rich BR (1952) Investigation of heat from inclined flat plate in free convection. ASME-meeting F-20 American society of mechanical engineers, NY, 52-F-20
22. Michiyoshi I (1964) Heat transfer from inclined thin flat plate by natural convection. Jpn Soc Mech Eng Bull 7(28):745–750
23. Kierkus WT (1968) An analysis of laminar free convection flow and heat transfer about an inclined isothermal plate. Int J Heat Mass Transf 11(2):241–253
24. Vliet GC (1969) Natural convection local heat transfer on constant-heat-flux inclined surfaces. J Heat Transf 91(4):511–516
25. Hassan KE, Mohamed SA (1970) Natural convection from isothermal flat surfaces. Int J Heat Mass Transf 13(12):1873–1886
26. Lee JB, Lock GSH (1972) Laminar boundary-layer free convection along an inclined, isothermal surface. Trans CSME 1:189–196
27. Fuji T, Imura H (1972) Natural-convection heat transfer from a plate with arbitrary inclination. Int J Heat Mass Transf 15(4):755–764
28. Vliet GC, Ross DC (1975) Turbulent natural convection on upward and downward facing inclined constant heat flux surfaces. J Heat Transf 97(4):549–554
29. Black WZ, Norris JK (1975) The thermal structure of free convection turbulence from inclined isothermal surfaces and its influence on heat transfer. Int J Heat Mass Transf 18(1):43–50

30. Gryzagoridis J, Klingenberg BE (1986) Natural convection from upper and lower surfaces of an inclined isothermal plate. Int Commun Heat Mass Transf 13(2):163–169
31. King JA, Reible DD (1991) Laminar natural convection heat transfer from inclined surfaces. Int J Heat Mass Transf 34(7):1901–1904
32. Wei JJ, Yu B, Wang HS, Tao WQ (2002) Numerical study of simultaneous natural convection heat transfer from both surfaces of a uniformly heated thin plate with arbitrary inclination. Heat Mass Transf/Waerme- und Stoffuebertragung 38(4–5):309–317
33. Al-Arabi M, Sakr B (1988) Natural convection heat transfer from inclined isothermal plates. Int J Heat Mass Transf 31(3):559–566
34. Jeschke P, Beer H (2001) Longitudinal vortices in a laminar natural convection boundary layer flow on an inclined flat plate and their influence on heat transfer. J Fluid Mech 432:313–339
35. Kimura F, Kitamura K, Yamaguchi M, Asami T (2003) Fluid flow and heat transfer of natural convection adjacent to upward-facing inclined heated plates. Heat Transf Asian Res 32(3):278–291
36. Kimura F, Yoshioka T, Kitamura K, Yamaguchi M, Asami T (2002) Fluid flow and heat transfer of natural convection at a slightly inclined, upward-facing, heated plate. Heat Transf Asian Res 31(5):362–375
37. Komori K, Kito S, Nakamura T, Inaguma Y, Inagaki T (2001) Fluid flow and heat transfer in the transition process of natural convection over an inclined plate. Heat Transf Asian Res 30(8):648–659
38. Lloyd JR, Sparrow EM (1970) On the instability of natural convection flow on inclined plates. J Fluid Mech 42:465–470
39. Sparrow EM, Husar RB (1969) Longitudinal vortices in natural convection flow on inclined plates. J Fluid Mech 37:251–255
40. Sparrow EM, Ramsey JW, Mass EA (1979) Effect of finite width on heat transfer and fluid flow about an inclined rectangular plate. J Heat Transf 101(2):199–204
41. Oosthuizen PH (1965) An experimental analysis of the heat transfer by laminar free convection heat transfer from a narrow vertical flat plate. JSA Inst Mech Eng 14(7):153–158
42. Oosthuizen PH (1967) A further experimental study of the laminar free convective heat transfer from narrow vertical plates in air. JSA Inst Mech Eng 16(9):182–184
43. Oosthuizen PH, Paul JT (1985) Numerical study of free convective heat transfer from narrow vertical flat plates. Proceedings of 10th Canadian congress applied mechanics. pp C23–C24
44. Oosthuizen PH, Henderson C (1987) Edge effects on forced and free convective laminar boundary layer flow over a flat plate. Proceedings 1987 ASME winter annual meeting HTD-82 convective transport. pp 149–155
45. Noto K, Matsumoto R (1984) Three-dimensional analysis of natural convection around a vertical plate of short height and narrow width. Nippon Kikai Gakkai Ronbunshu, B Hen/Trans Jpn Soc Mech Eng B 50(453):1431–1436
46. Noto K, Matsumoto R (1985) Three-dimensional analysis of the Navier-Stokes equations on laminar natural convection around a vertical flat plate. Proceedings 4th international conference numerical methods in laminar and turbulent flow. pp 865–877
47. Noto K, Matsumoto R (1987) Three-dimensional numerical analysis of natural convective heat transfer from a vertical plate. Proceedings 1987 ASME/JSME thermal engineering joint conference. pp 1–8
48. Oosthuizen PH, Paul JT (2006) Natural convective heat transfer from a narrow isothermal vertical flat plate. Proceedings 9th AIAA/ASME joint thermophysics heat transfer conference paper AIAA 2006. p 3397
49. Oosthuizen PH, Paul JT (2007) Natural convective heat transfer from a narrow vertical flat plate with a uniform heat flux at the surface. Proceedings 2007 ASME/JSME thermal engineering summer heat transfer conference paper HT2007. p 32134
50. Kalendar AY, Oosthuizen PH (2008) Natural convective heat transfer from an inclined narrow isothermal flat plate. Proceedings ASME National heat transfer conference paper HT2008. p 56190

51. Kalendar AY, Oosthuizen PH (2011) Numerical and experimental studies of natural convective heat transfer from vertical and inclined narrow isothermal flat plates. Heat Mass Transf 47(9):1181–1195
52. Kalendar AY, Oosthuizen PH (2010) A numerical study of natural convective heat transfer from vertical and inclined narrow isothermal flat plates in the transition and turbulent flow regions. Proceedings of 2010 CSME Forum
53. Oosthuizen PH, Paul JT (2007) Effect of edge conditions on natural convective heat transfer from a narrow vertical flat plate with a uniform surface heat flux. Proceedings ASME international mechanical engineering congress and exposition paper IMECE2007-42712. pp 397–404
54. Oosthuizen PH, Paul JT (2007) Natural convective heat transfer from a narrow vertical isothermal flat plate with different edge conditions. Proceedings of 15th annual CFD society Canada conference, Toronto
55. Oosthuizen PH, Paul JT (2010) Natural convective heat transfer from a narrow vertical flat plate with a uniform surface heat flux and with different plate edge conditions. Front Heat Mass Transf. doi:10.5098/hmt.v1.1.3006
56. Oosthuizen PH, Paul JT (2007) Natural convective heat transfer from a recessed narrow vertical flat plate with a uniform heat flux at the surface. Proceedings of the 5th international conference heat transfer, fluid mechanics and thermodynamics (HEFAT2007), Sun City, SA, 1–4 July, p 6
57. Baker E (1972) Liquid cooling of microelectronic devices by free and forced convection. Microelectron Reliab 11(2):213–222
58. Baker E (1973) Liquid immersion cooling of small electronic devices. Microelectron Reliab 12(2):163–173
59. Park KA, Bergles AE (1987) Natural convection heat transfer characteristics of simulated microelectronic chips. J Heat Transf 109(1):90–96
60. Kalendar AY, Oosthuizen PH (2011) Numerical and experimental studies of natural convective heat transfer from vertical and inclined narrow isothermal flat plates. Heat Mass Transf 47(9):1181–1195
61. Kalendar AY, Oosthuizen PH (2010) A numerical study of natural convective heat transfer from vertical and inclined narrow isothermal flat plates in the transition and turbulent flow regions. Proceedings of 2010 CSME Forum, Victoria, BC
62. Heindel TJ, Incropera FP, Ramadhyani S (1996) Enhancement of natural convection heat transfer from an array of discrete heat sources. Int J Heat Mass Transf 39(3):479–490
63. Tou SKW, Tso CP, Zhang TX (1999) 3-D numerical analysis of natural convective liquid cooling of a 33 heater array in rectangular enclosures. Int J Heat Mass Transf 42(17):3231–3244
64. Tou SKW, Zhang XF (2003) Three-dimensional numerical simulation of natural convection in an inclined liquid-filled enclosure with an array of discrete heaters. Int J Heat Mass Transf 46(1):127–138
65. Baskaya S, Erturhan U, Sivrioglu M (2005) An experimental study on convection heat transfer from an array of discrete heat sources. Int Commun Heat Mass Transf 32(1–2):248–257
66. Oosthuizen PH, Paul JT (2005) Natural convection in a rectangular enclosure with two heated sections on the lower surface. Int J Heat Fluid Flow 26(4):587–596
67. Kirby P, Fleischer A (2007) Thermal interaction in free convection from two unequally powered discrete heat sources with various orientations and separation distances on a vertical plate. Proceedings of ASME international mechanical engineering congress and exposition, Paper MECE2007-44053
68. Kalendar AY, Oosthuizen PH (2010) Natural convective heat transfer from two adjacent isothermal narrow vertical and inclined flat plates. JP J Heat Mass Transf 4(1):61–80
69. Kalendar AY, Oosthuizen PH (2009) Natural convective heat transfer from two vertically spaced narrow isothermal flat plates. Proceedings of 17th annual conference CFD society, Canada

70. Kalendar AY, Oosthuizen PH, Kalandar B (2009) A numerical study of natural convective heat transfer from two adjacent narrow isothermal inclined flat plates. Proceedings ASME 2009 heat transfer summer conference (HT2009), California. Paper HT2009-88091. doi: http://dx.doi.org/10.1115/HT2009-88091

Chapter 2
Natural Convective Heat Transfer from Narrow Vertical Plates

Keywords Natural convection · Narrow plates · Vertical plates · Isothermal plates · Uniform surface heat flux · Numerical · Empirical equations

2.1 Introduction

As discussed in the preceding chapter, two-dimensional natural convective heat transfer from wide vertical plates has been extensively studied and many equations for predicting the mean and local heat transfer rates from such plates have been derived. However, when the width of the plate is relatively small compared to its height, i.e., when the plate is narrow, the heat transfer rate can be considerably greater than that predicted by these wide plate equations. As explained in Chap. 1, the increase in the heat transfer rate from a narrow plate relative to that from a wide plate under the same conditions results from the fact that fluid flow is induced inwards near the edges of the plate and the flow near the edge of the plate is, thus, three-dimensional. With a wide plate the effects of this edge flow are negligible but with a narrow plate these effects can be very significant. The increase in the heat transfer rate is often, therefore, said to be due to "three-dimensional edge effects" or just "edge effects". Situations that can be approximately modeled as narrow vertical plates occur in a number of practical situations, so there exists a need to be able to predict heat transfer rates from such narrow plates.

In the present chapter, numerically determined natural convective heat transfer rates from narrow vertical plates for a relatively wide range of dimensionless plate widths will be discussed. Attention has here been restricted to plates having a uniform surface temperature and to plates having a uniform surface heat flux. Results will be presented for a Prandtl number of 0.7, this being approximately the value for air at ambient conditions and the results thus apply to the conditions existing in many applications in which the present results are relevant.

P. H. Oosthuizen and A. Y. Kalendar, *Natural Convective Heat Transfer from Narrow Plates*, SpringerBriefs in Thermal Engineering and Applied Science, DOI: 10.1007/978-1-4614-5158-7_2, © The Author(s) 2013

Fig. 2.1 Flow situation
considered

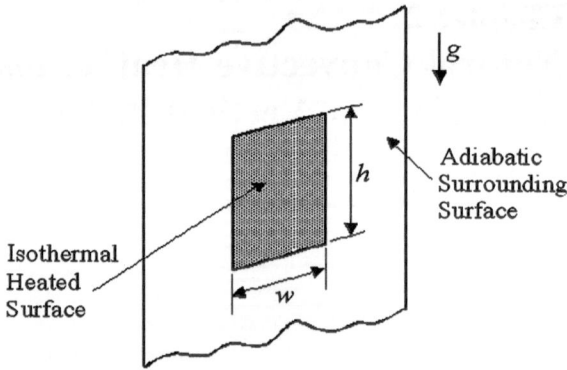

Previous studies of natural convective heat transfer from narrow vertical plates
were discussed in Chap. 1. The work described in this chapter is based on the
studies reported by Oosthuizen and Paul [1, 2].

2.2 Isothermal Plates

Attention will first be given to the case where the surface of the plate is at a
uniform temperature, i.e., where the plate is isothermal. The flow situation considered is shown in Fig. 2.1. The vertical isothermal plate, as shown in this figure,
is surrounded by an adiabatic surface in the same plane as the heated plate. The
width of the plate, w, is assumed to be less than the vertical height of the plate, h.

2.2.1 Solution Procedure

In obtaining the results discussed here, the flow has been assumed to be laminar
and it has been assumed that the fluid properties are constant except for the density
change with temperature which gives rise to the buoyancy forces, this having been
treated by using the Boussinesq approach. It has also been assumed that the flow is
symmetric about the vertical center-plane of the plate. The solution has been
obtained by numerically solving the full three-dimensional form of the governing
equations, these equations being written in terms of dimensionless variables using
the height, h, of the heated plate as the length scale and the overall temperature
difference $(T_H - T_F)$ as the temperature scale, T_F being the fluid temperature far
from the plate, and T_H being the uniform surface temperature of the plate.

Defining the following reference velocity:

$$u_r = \frac{\alpha}{h}\sqrt{Ra\,Pr} \tag{2.1}$$

Fig. 2.2 Solution domain
ABCDEFIJLM

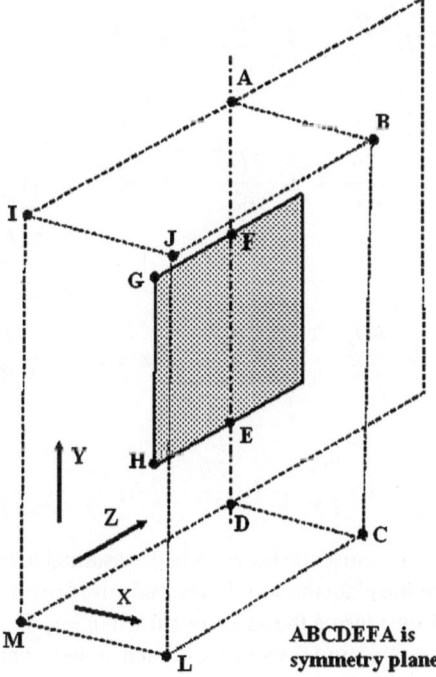

where Pr is the Prandtl number and Ra is the Rayleigh number based on h, i.e.,:

$$Ra = \frac{\beta g(T_H - T_F)h^3}{\nu\alpha} \tag{2.2}$$

the following dimensionless variables have then been defined:

$$X = \frac{x}{h}, \; Y = \frac{y}{h}, \; Z = \frac{z}{h}, \; U_X = \frac{u_x}{u_r}, \; U_Y = \frac{u_y}{u_r},$$

$$U_Z = \frac{u_z}{u_r}, \; P = \frac{(p - p_F)h}{\mu u_r}, \; \theta = \frac{T - T_F}{T_H - T_F} \tag{2.3}$$

where T is the temperature. The X-coordinate is measured in the horizontal direction normal to the plate, the Y-coordinate is measured in the vertically upward direction and the Z-coordinate is measured in the horizontal direction in the plane of plate as shown in Fig. 2.2.

In terms of these dimensionless variables, the governing equations are:

$$\frac{\partial U_X}{\partial X} + \frac{\partial U_Y}{\partial Y} + \frac{\partial U_Z}{\partial Z} = 0 \tag{2.4}$$

$$U_X \frac{\partial U_X}{\partial X} + U_Y \frac{\partial U_X}{\partial Y} + U_Z \frac{\partial U_z}{\partial Z} = \sqrt{\frac{Pr}{Ra}} \left(-\frac{\partial P}{\partial X} + \frac{\partial^2 U_X}{\partial X^2} + \frac{\partial^2 U_X}{\partial Y^2} + \frac{\partial^2 U_X}{\partial Z^2} \right)$$

(2.5)

$$U_X \frac{\partial U_Y}{\partial X} + U_Y \frac{\partial U_Y}{\partial X} + U_Z \frac{\partial U_Y}{\partial X} = \sqrt{\frac{Pr}{Ra}} \left(-\frac{\partial P}{\partial Y} + \frac{\partial^2 U_Y}{\partial X^2} + \frac{\partial^2 U_Y}{\partial Y^2} + \frac{\partial^2 U_Y}{\partial Z^2} \right)$$
$$+ T$$

(2.6)

$$U_X \frac{\partial U_Z}{\partial X} + U_Y \frac{\partial U_Z}{\partial Y} + U_Z \frac{\partial U_Z}{\partial Z} = \sqrt{\frac{Pr}{Ra}} \left(-\frac{\partial P}{\partial Z} + \frac{\partial^2 U_Z}{\partial X^2} + \frac{\partial^2 U_Z}{\partial Y^2} + \frac{\partial^2 U_Z}{\partial Z^2} \right)$$

(2.7)

$$U_X \frac{\partial \theta}{\partial X} + U_Y \frac{\partial \theta}{\partial Y} + U_Z \frac{\partial \theta}{\partial Z} = \frac{1}{\sqrt{RaPr}} \left(\frac{\partial^2 \theta}{\partial X^2} + \frac{\partial^2 \theta}{\partial Y^2} + \frac{\partial^2 \theta}{\partial Z^2} \right) \qquad (2.8)$$

Because the flow has been assumed to be symmetric about the vertical center-line of the plate the solution domain used in obtaining the solution is as shown in Fig. 2.2. Considering the surfaces shown in Fig. 2.2, the assumed boundary conditions on the solution in terms of the dimensionless variables are, since flow symmetry is being assumed:

$$\begin{aligned} \text{FEHG:} \quad & U_X = 0, \ U_Y = 0, \ U_Z = 0, \theta = 1 \\ \text{AFEDMI except for FEHG:} \quad & U_X = 0, \ U_Y = 0, \ U_Z = 0, \frac{\partial \theta}{\partial X} = 0 \\ \text{IJLM:} \quad & U_Y = 0, \ U_X = 0, \ \theta = 0 \\ \text{DCLM:} \quad & U_X = 0, \ U_Z = 0, \ \theta = 0 \\ \text{BCLJ:} \quad & U_Y = 0, \ U_Z = 0, \ \theta = 0 \\ \text{ABCDEF:} \quad & U_Z = 0, \frac{\partial U_Y}{\partial Z} = 0, \frac{\partial U_X}{\partial Z} = 0, \frac{\partial \theta}{\partial Z} = 0 \end{aligned}$$

The mean heat transfer rate from the heated plate has been expressed in terms of the following mean Nusselt number:

$$Nu = \frac{\bar{q}'}{k(T_H - T_F)} \qquad (2.9)$$

The above dimensionless governing equations subject to the boundary conditions listed above have been numerically solved using the commercial CFD solver FLUENT©. Extensive grid and convergence criterion independence testing was undertaken. This indicated that the heat transfer results presented here are to within 1 % independent of the number of grid points and of the convergence criterion used. The effect of the positioning of the outer surfaces of the solution domain (i.e., surfaces MLCD, IJBA, IJLM, and BCLJ in Fig. 2.2) from the heated surface

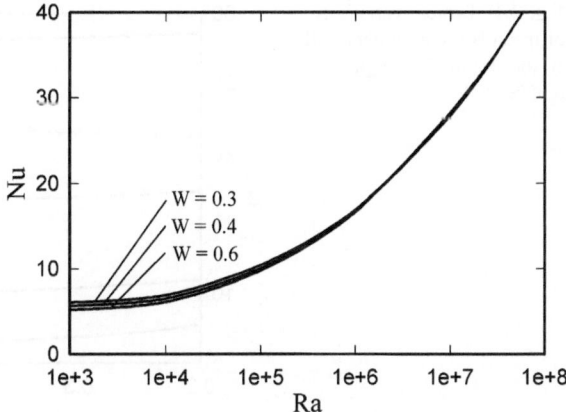

Fig. 2.3 Typical variations of mean Nusselt number with Rayleigh number for various W values

was also examined and the positions used in obtaining the results discussed here were chosen to ensure that the heat transfer results were independent of this positioning to within 1 %.

2.2.2 Results

The solution has the following parameters:

1. The Rayleigh number, Ra, based on the plate height, h, and the overall temperature difference between the plate temperature and the fluid temperature.
2. The dimensionless plate width, $W = w/h$.
3. The Prandtl number, Pr.

As previously mentioned, results have been obtained only for $Pr = 0.7$. Ra values between 10^2 and 10^8 and W values between 0.2 and 0.6 have been considered. Attention has therefore been restricted to laminar flow over the plates. The effect of the development of turbulent flow will be discussed in a later chapter.

Typical variations of the mean Nusselt number for the plate with Rayleigh number for various dimensionless plate widths are shown in Fig. 2.3. It will be seen from these results, particularly at the lower Rayleigh numbers, that the mean Nusselt number tends to increase with decreasing W. This is further illustrated by the results shown in Fig. 2.4. It will be seen from the results in Fig. 2.4 that for $Ra = 10^4$, the mean Nusselt increases by approximately 25 % as W decreases from 0.6 to 0.2. For $Ra = 10^7$, however, the increase is less than 2 %.

As mentioned previously, the increase in the Nusselt number with decreasing W arises from the fact that there is an induced inflow toward the plate from the sides and this causes the heat transfer rate to be higher near the vertical edges of the plate than it is in the center region of the plate. This edge effect is illustrated by the results given in Figs. 2.5 and 2.6 which show the local heat transfer rate

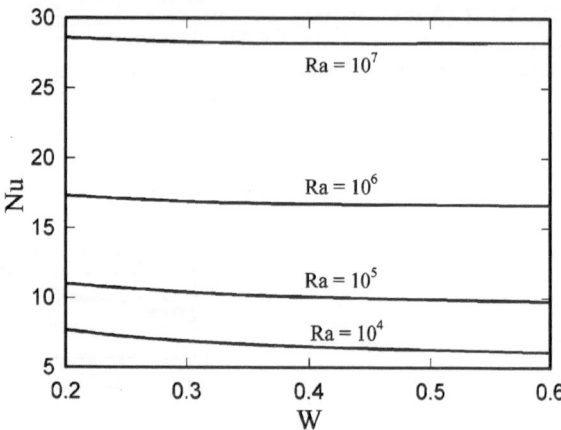

Fig. 2.4 Typical variations of mean Nusselt number with W for various Rayleigh number values

distributions over the plate surface. The results given in these two figures illustrate how the edge effect grows more significant as W is decreased and how the extent of the edge regions increases compared to the overall plate area. Since the extent of the edge region will also be a function of the boundary layer thickness which increases as the Rayleigh number decreases, the importance of the edge effect increases as the Rayleigh number decreases.

In order to obtain an approximate equation for the Nusselt number for narrow plates, it has been assumed that for wide plates, i.e., for situations in which the edge effects are negligible, Nu is approximately given by the widely used Churchill and Chu correlation [3], that is by:

$$Nu_0 = 0.68 + \left\{ \frac{0.67}{\left[1 + (0.492/\,Pr)^{9/16}\right]^{4/9}} \right\} Ra^{0.25} \qquad (2.10)$$

the subscript 0 indicating that this value of the Nusselt number applies to a wide plate. This equation gives for $Pr = 0.7$:

$$Nu_0 = 0.68 + 0.513 Ra^{0.25} \qquad (2.11)$$

A comparison between the results given by this equation and the numerical results for $W = 0.6$ at the higher Rayleigh numbers considered is shown in Fig. 2.7. It will be seen from Fig. 2.7 that there is good agreement between the numerical results and the correlation equation for Rayleigh numbers above 10^4. At the lower Rayleigh numbers, the numerical results are higher than the correlation equation results due to edge effects.

Because of the assumed form of the variation of Nu when edge effects are negligible, it has been assumed that the Nusselt number for a plate of dimensionless width W is given by an equation of the form:

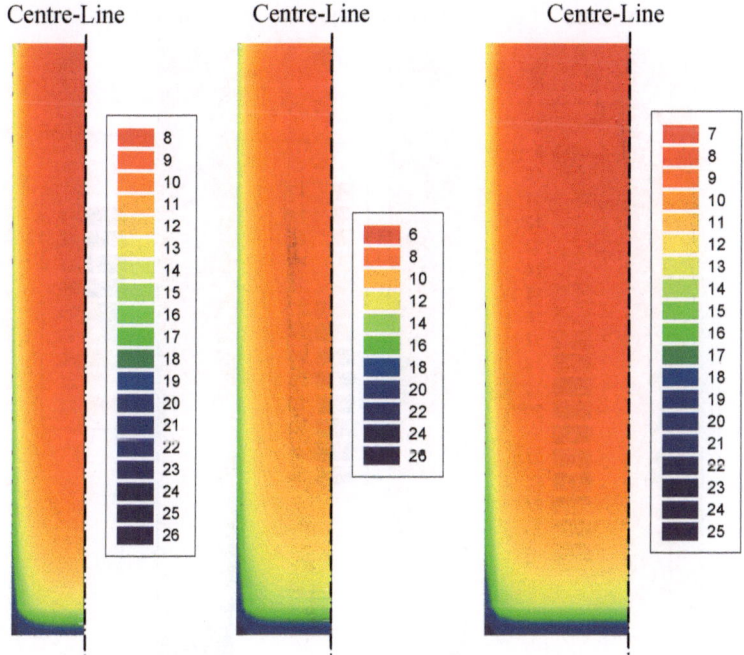

Fig. 2.5 Local Nusselt number (based on h) distributions over plate for $Ra = 10^5$ for $W = 0.25$ (*left*), 0.35 (*center*) and 0.5 (*right*)

$$\frac{Nu}{Ra^{0.25}} = \frac{Nu_0}{Ra^{0.25}} + \text{function}\,(Ra,\,W) \tag{2.12}$$

The function of Ra and W was determined from the numerical results. It was found that the edge effects are dependent on the value of $WRa^{0.5}$ and that the heat transfer results obtained numerically can be approximately described by:

$$\frac{Nu}{Ra^{0.25}} = \frac{Nu_0}{Ra^{0.25}} + \frac{7.1}{(WRa^{0.5})^{1.25}} \tag{2.13}$$

A comparison of the results given by this equation and some of the numerical results is shown in Fig. 2.8. It will be seen that the equation describes 95 % of the computed results to an accuracy of better than approximately 5 %.

If three-dimensional effects (i.e., edge effects) are assumed to be negligible when:

$$\frac{Nu/Ra^{0.25} - Nu_0/Ra^{0.25}}{Nu_0/Ra^{0.25}} < 0.01 \tag{2.14}$$

then, since $Nu_0/Ra^{0.25}$ is approximately equal to 0.5, the above equation gives the following approximate criterion for when three-dimensional edge effects are negligible:

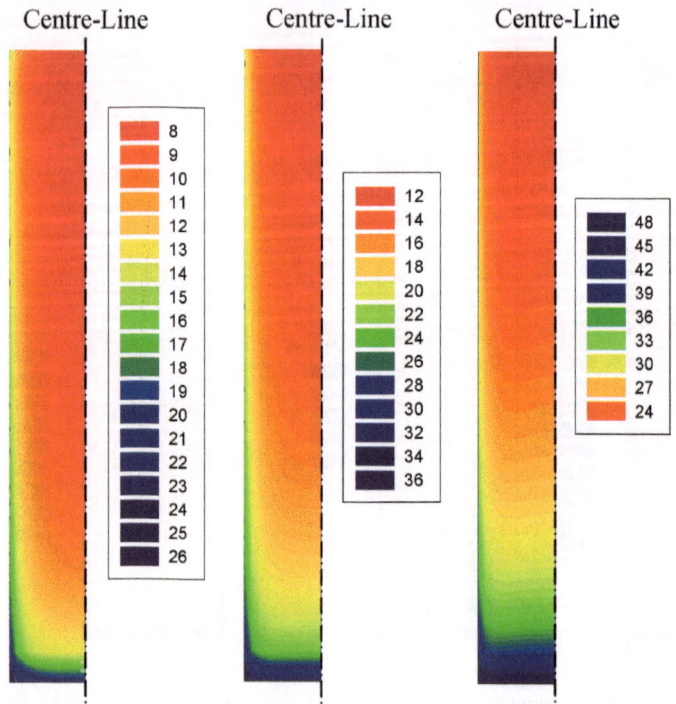

Fig. 2.6 Local Nusselt number (based on h) distributions over plate for $W = 0.25$ for $Ra = 10^5$ (*left*), $Ra = 10^6$ (*center*) and $Ra = 10^7$ (*right*)

Fig. 2.7 Comparison between numerical results for $W = 0.6$ and the Churchill and Chu correlation equation for wide vertical plates

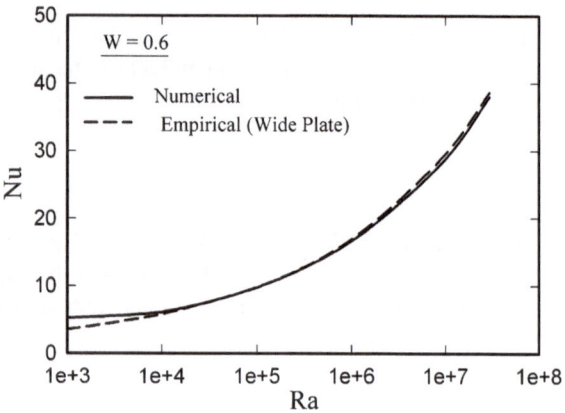

$$\frac{Nu}{Ra^{0.25}} - \frac{Nu_0}{Ra^{0.25}} < 0.005 \qquad (2.15)$$

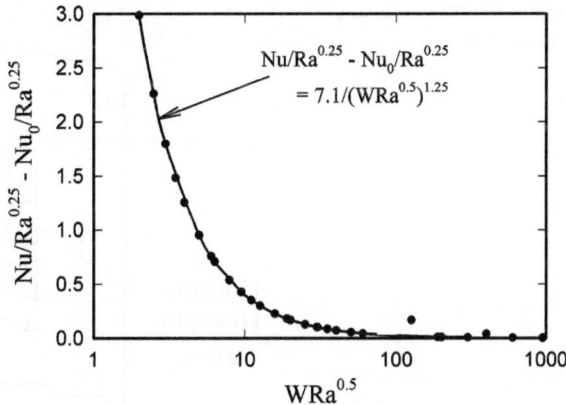

Fig. 2.8 Comparison of correlation for narrow plates with numerical results

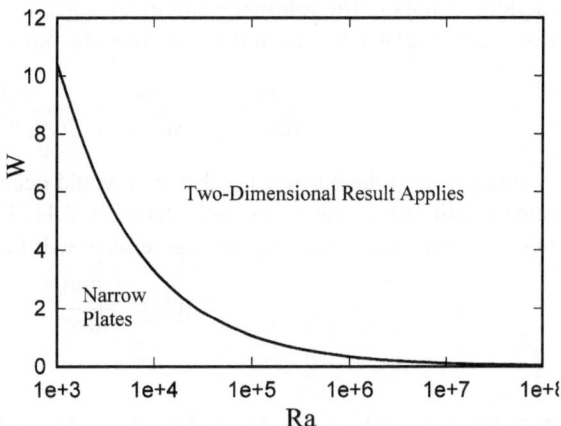

Fig. 2.9 Variation of W value above which edge effects are negligible with Ra

i.e., using Eq. 2.13 when:

$$\frac{7.1}{(WRa^{0.5})^{1.25}} < 0.005, \quad \text{i.e.,} \quad WRa^{0.5} > 330, \quad \text{i.e.,} \quad W > \frac{330}{Ra^{0.5}} \quad (2.16)$$

Using this result, the variation of the value of W above which three-dimensional edge effects on Nu are negligible (to within approximately 1 %) with Ra can be determined and is shown in Fig. 2.9.

2.2.3 Concluding Remarks

The dimensionless plate width has been shown to have a significant influence on the mean Nusselt number for natural convective heat transfer from a vertical isothermal flat plate. This effect has been shown to increase with decreasing

Fig. 2.10 Flow situation
considered

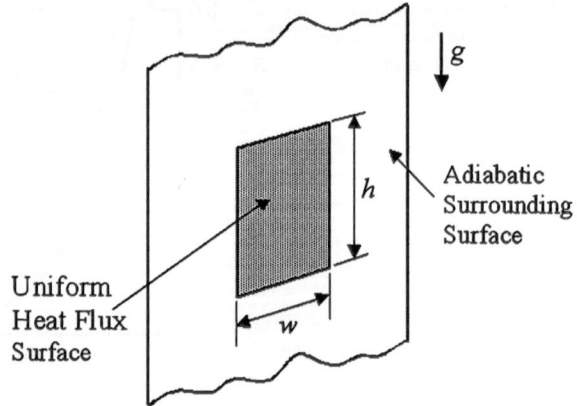

Rayleigh number. The following empirical equation for the mean heat transfer rate
from narrow plates has been derived from the numerical results.

$$\frac{Nu}{Ra^{0.25}} = \frac{Nu_0}{Ra^{0.25}} + \frac{7.1}{\left(WRa^{0.5}\right)^{1.25}}$$

where Nu_0 is the Nusselt number that would occur with a wide plate under the
same conditions, its value being given by Eq. 2.11. The above correlation equation
indicates that three-dimensional edge effects will be negligible if:

$$W > \frac{330}{Ra^{0.5}}$$

2.3 Plates with a Uniform Surface Heat Flux

Attention will next be given to the case where the plate has a uniform heat flux
over its surface. The flow situation considered is shown in Fig. 2.10. The vertical
plate is, as shown in this figure, again surrounded by an adiabatic surface in the
same plane as the heated plate. The width of the plate, w, is again assumed to be
less than the vertical height of the plate, h.

2.3.1 Solution Procedure

The same basic assumptions that were used in dealing with the isothermal plate
have been adopted, i.e., the flow has again been assumed to be steady and laminar
and it has again been assumed that the fluid properties are constant except for the
density change with temperature which gives rise to the buoyancy forces, this
having been treated by using the Boussinesq approach. It has also been assumed
that the flow is symmetric about the vertical center-plane of the plate and the

solution has again been obtained by numerically solving the full three-dimensional form of the governing equations, these equations being written in terms of dimensionless variables using the height, h, of the heated plate as the length scale and $q'_w h/k$ as the temperature scale. Defining the following reference velocity:

$$u_r = \frac{\alpha}{h}\sqrt{Ra^* Pr} \tag{2.17}$$

where Pr is the Prandtl number and Ra^* is the heat flux Rayleigh number based on h, i.e.,:

$$Ra^* = \frac{\beta g q'_w h^4}{k\nu\alpha} \tag{2.18}$$

the following dimensionless variables have then been introduced:

$$X = \frac{x}{h}, \; Y = \frac{y}{h}, \; Z = \frac{z}{h}, \; U_X = \frac{u_x}{u_r}, \; U_Y = \frac{u_y}{u_r},$$

$$U_Z = \frac{u_z}{u_r}, \; P = \frac{(p - p_F)h}{\mu u_r}, \; \theta = \frac{T - T_F}{q'_w h/k} \tag{2.19}$$

where T, is the temperature and T_F is the fluid temperature far from the plate. As with the isothermal plate, the X-coordinate is measured in the horizontal direction normal to the plate, the Y-coordinate is measured in the vertically upward direction and the Z-coordinate is measured in the horizontal direction in the plane of plate.

In terms of these dimensionless variables, the governing equations are:

$$\frac{\partial U_X}{\partial X} + \frac{\partial U_Y}{\partial Y} + \frac{\partial U_Z}{\partial Z} = 0 \tag{2.20}$$

$$U_X\frac{\partial U_X}{\partial X} + U_Y\frac{\partial U_X}{\partial Y} + U_Z\frac{\partial U_X}{\partial Z} = \sqrt{\frac{Pr}{Ra^*}}\left(-\frac{\partial P}{\partial X} + \frac{\partial^2 U_X}{\partial X^2} + \frac{\partial^2 U_X}{\partial Y^2} + \frac{\partial^2 U_X}{\partial Z^2}\right) \tag{2.21}$$

$$U_X\frac{\partial U_Y}{\partial X} + U_Y\frac{\partial U_Y}{\partial X} + U_Z\frac{\partial U_Y}{\partial X} = \sqrt{\frac{Pr}{Ra^*}}\left(-\frac{\partial P}{\partial Y} + \frac{\partial^2 U_Y}{\partial X^2} + \frac{\partial^2 U_Y}{\partial Y^2} + \frac{\partial^2 U_Y}{\partial Z^2}\right) + T \tag{2.22}$$

$$U_X\frac{\partial U_Z}{\partial X} + U_Y\frac{\partial U_Z}{\partial Y} + U_Z\frac{\partial U_Z}{\partial Z} = \sqrt{\frac{Pr}{Ra^*}}\left(-\frac{\partial P}{\partial Z} + \frac{\partial^2 U_Z}{\partial X^2} + \frac{\partial^2 U_Z}{\partial Y^2} + \frac{\partial^2 U_Z}{\partial Z^2}\right) \tag{2.23}$$

$$U_X\frac{\partial \theta}{\partial X} + U_Y\frac{\partial \theta}{\partial Y} + U_Z\frac{\partial \theta}{\partial Z} = \frac{1}{\sqrt{Ra^* Pr}}\left(\frac{\partial^2 \theta}{\partial X^2} + \frac{\partial^2 \theta}{\partial Y^2} + \frac{\partial^2 \theta}{\partial Z^2}\right) \tag{2.24}$$

Because the flow has been assumed to be symmetric about the vertical center-line of the plate the solution domain used in obtaining the solution is as shown in

Fig. 2.11 Solution domain
ABCDIJLM

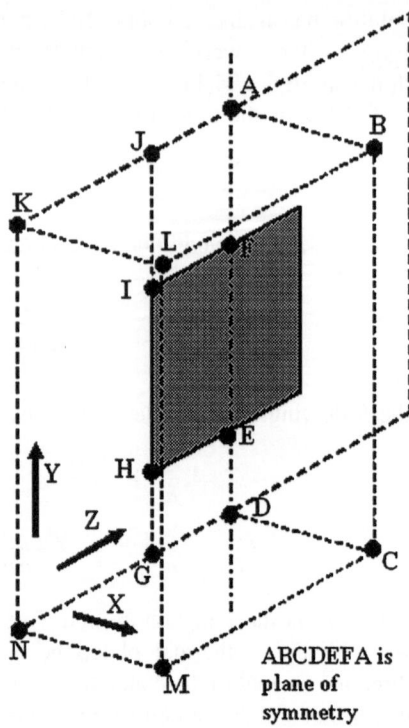

ABCDEFA is
plane of
symmetry

Fig. 2.11. Considering the surfaces shown in Fig. 2.11, the assumed boundary conditions on the solution in terms of the dimensionless variables are, since flow symmetry is being assumed:

$$\text{FEHIF:}\quad U_X = 0,\ U_Y = 0,\ U_Z = 0,\ \frac{\partial\theta}{\partial X} = -1$$

$$\text{AFEDGNKJA except for FEHIF:}\quad U_X = 0,\ U_Y = 0,\ U_Z = 0,\ \frac{\partial\theta}{\partial X} = 0$$

$$\text{KLMNK:}\quad U_Y = 0,\ U_X = 0,\ \theta = 0$$

$$\text{CMNDC:}\quad U_X = 0,\ U_Z = 0,\ \theta = 0$$

$$\text{BCMLB:}\quad U_Y = 0,\ U_Z = 0,\ \theta = 0$$

$$\text{ABCDEFA:}\quad U_Z = 0,\ \frac{\partial U_Y}{\partial Z} = 0,\ \frac{\partial U_X}{\partial Z} = 0,\ \frac{\partial\theta}{\partial Z} = 0$$

The mean Nusselt number for the heated plate is defined by:

$$Nu = \frac{q'_w h}{k(\bar{T}_H - T_F)} \tag{2.25}$$

where $\bar{T}_H$ is the mean surface temperature of the plate.

The above dimensionless governing equations subject to the boundary conditions listed above have been numerically solved using a commercial CFD code. Extensive grid and convergence criterion independence testing was again

Fig. 2.12 Typical variations of mean Nusselt number with Rayleigh number for various W values

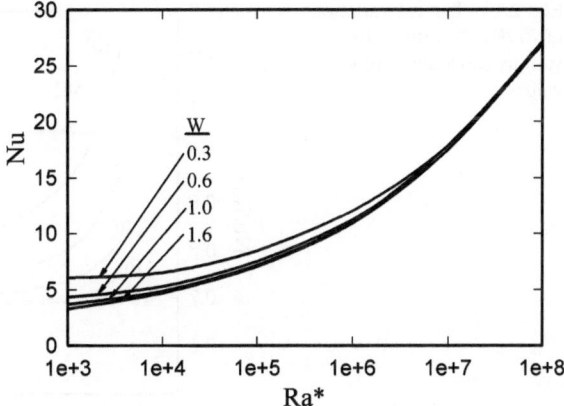

undertaken. This indicated that the heat transfer results presented here are to within 1 % independent of the number of grid points used and of the convergence criterion used. The effect of the positioning of the outer surfaces of the solution domain (i.e., surfaces MLCD, IJBA, IJLM, and BCLJ in Fig. 2.11) from the heated surface was also examined and the positions used in obtaining the results discussed here were again chosen to ensure that the heat transfer results were independent of this positioning to within 1 %.

2.3.2 Results

The solution has the following parameters:

1. The heat flux Rayleigh number, Ra^*,
2. The dimensionless plate width, $W = w/h$.
3. The Prandtl number, Pr.

As already mentioned, results have again only been obtained for $Pr = 0.7$. Ra values between 10^2 and 10^8 and W values between 0.2 and 1.6 have been considered.

Typical variations of the mean Nusselt number for the plate with Rayleigh number for various dimensionless plate widths are shown in Fig. 2.12. It will be seen from these results that, particularly at the lower Rayleigh numbers, the mean Nusselt number tends to increase with decreasing W. This is further illustrated by the results shown in Fig. 2.13. In this figure, it has been noted that with two-dimensional flow over a wide vertical plate that has a uniform heat flux over its surface:

$$\frac{Nu}{Ra^{*0.2}} = \text{function}(Pr) \tag{2.26}$$

Fig. 2.13 Typical variations of $Nu/Ra^{*0.2}$ with W for various Rayleigh number values

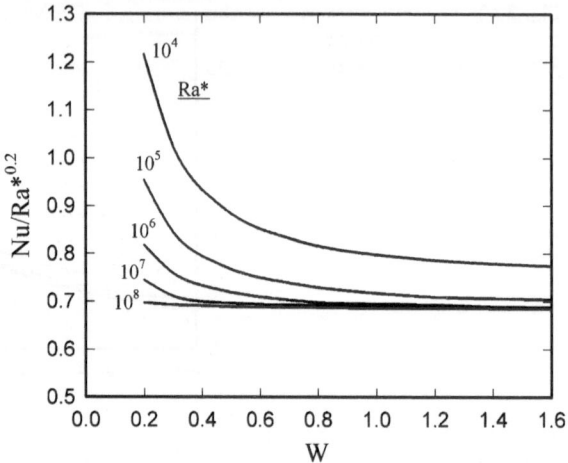

Figure 2.13 therefore shows the variation of $Nu/Ra^{*0.2}$ with the dimensionless plate width, W, for various values of Ra^{*}. It will be seen that at the larger values of W and Ra^{*} the quantity $Nu/Ra^{*0.2}$ does tend to a constant value of approximately 0.68 which is the value given by the similarity solution for two-dimensional flow over a wide plate with a uniform heat flux at the surface for a Prandtl number of 0.7, e.g., see Kakac and Yener [4]. It will also be seen from Fig. 2.13 that, at the lowest values of W and Ra^{*} considered, the value of $Nu/Ra^{*0.2}$ is more than 70 % above the two-dimensional flow value. However, at the highest value of Ra^{*} considered the two-dimensional value applies at W values down to about 0.3. As mentioned before, the increase in the Nusselt number with decreasing W arises from the fact that there is an induced inflow toward near the edges of the plate which causes the heat transfer rate to be higher near the vertical edges of the plate than it is in the center region of the plate.

In order to obtain an approximate equation for the Nusselt number for narrow plates with a uniform surface heat flux, it has been assumed that for wide plates, i.e., for situations in which the edge effects are negligible, Nu is given for $Pr = 0.7$, by the following equation (see, for example, Kakac and Yener [4]):

$$Nu = 0.68 \, Ra^{*0.2} \tag{2.27}$$

As discussed above, Fig. 2.13 shows that this equation gives results that are in good agreement with the present numerical results at the larger values of W and Ra^{*} considered. Because of this assumed form of the variation of Nu with Ra^{*} when edge effects are negligible, it has been assumed that the Nusselt number for a plate of arbitrary dimensionless width W is given by an equation of the form:

$$\frac{Nu}{Ra^{*0.2}} = 0.68 + \text{function}(Ra^{*}, W) \tag{2.28}$$

Fig. 2.14 Comparison of
correlation equation for
narrow plates with numerical
results

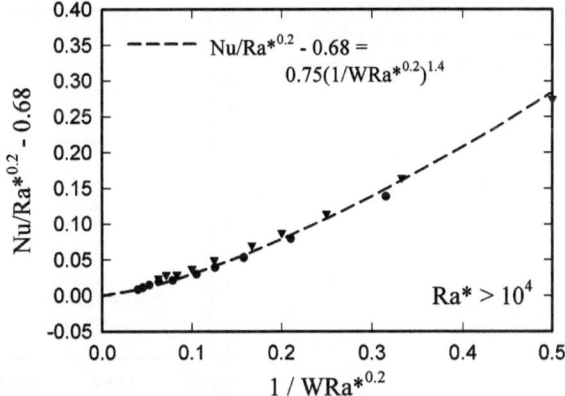

The function of Ra^* and W was determined from the numerical results which
indicated that the Nusselt number results obtained numerically could be approximately described by:

$$\frac{Nu}{Ra^{*0.2}} = 0.68 + \frac{0.75}{(WRa^{*0.2})^{1.4}} \qquad (2.29)$$

A comparison of the results given by this equation and the numerical results is
shown in Fig. 2.14. The quantity $WRa^{*0.2}$ is basically a measure of the ratio of the
plate width to the boundary layer thickness. When this ratio is large, the extent of
the plate area that is affected by three-dimensional effects is small and three-
dimensional flow (edge) effects are negligible.

If three-dimensional effects (i.e., edge effects) are assumed to be negligible if:

$$\frac{Nu/Ra^{*0.2} - Nu_0/Ra^{*0.2}}{Nu_0/Ra^{*0.2}} = \frac{Nu/Ra^{*0.2} - 0.68}{0.68} < 0.02 \qquad (2.30)$$

then the above equation gives the following approximate criterion of when three-
dimensional effects are negligible:

$$\frac{Nu}{Ra^{*0.2}} - 0.68 < 0.014 \qquad (2.31)$$

i.e., using Eq. 2.5:

$$\frac{0.75}{(WRa^{*0.2})^{1.4}} < 0.014, \quad \text{i.e.,} \quad WRa^{*0.2} > 15.2, \quad \text{i.e.,} \quad W > \frac{15.2}{Ra^{*0.2}} \qquad (2.32)$$

Using this result, the variation of the value of W above which three-dimensional
edge effects on Nu are negligible (to within approximately 2 %) with Ra^* can be
determined and is shown in Fig. 2.15.

Fig. 2.15 Variation of *W* value above which edge effects are negligible with *Ra*

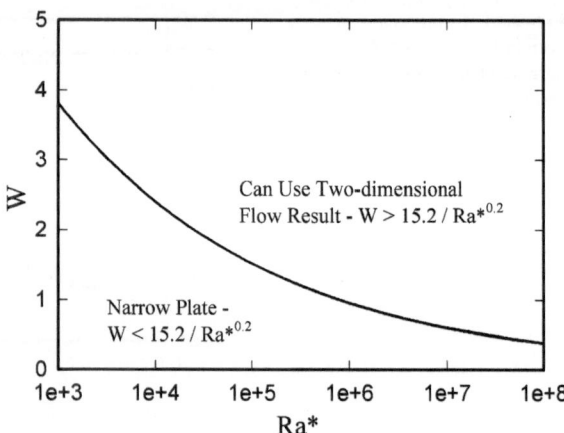

2.3.3 Conclusions

The dimensionless plate width has been shown to also have a significant influence on the mean Nusselt number for natural convective heat transfer from a vertical flat plate with a uniform heat flux at the surface. This effect has been shown to increase with decreasing dimensionless plate width and decreasing Rayleigh number. The following empirical equation for the mean heat transfer rate from narrow plates has been derived from the numerical results:

$$\frac{Nu}{Ra^{*0.2}} = 0.68 + \frac{0.75}{\left(WRa^{*0.2}\right)^{1.4}}$$

This equation indicates that three-dimensional effects will be negligible if:

$$W > \frac{15.2}{Ra^{*0.2}}$$

References

1. Oosthuizen PH, Paul JT (2006) Natural convective heat transfer from a narrow isothermal vertical flat plate. Proceedings of the 9th AIAA/ASME joint thermophysics and heat transfer conference, Paper AIAA 2006 3397
2. Oosthuizen PH, Paul JT (2007) Natural convective heat transfer from a narrow vertical flat plate with a uniform heat flux at the surface. Proceedings 2007 ASME-JSME thermal engineering summer heat transfer conference, Paper HT2007-32134
3. Churchill SW, Chu HHS (1975) Correlating equations for laminar and turbulent free convection from a vertical plate. Int J Heat Mass Transf 18(11):1323–1329
4. Kakac S, Yener Y (1995) Convective heat transfer, 2nd edn. CRC Press LLC, Boca Raton

Chapter 3
Natural Convective Heat Transfer from Inclined Narrow Plates

Keywords Natural convection · Narrow plates · Inclined plates · Isothermal plates · Numerical · Empirical equations

3.1 Introduction

In the preceding chapter, discussion was restricted to vertical plates. There are, however, a number of practical situations in which the heat transfer is effectively from an inclined flat plate. Therefore, in this chapter, numerical studies of natural convective heat transfer from an inclined narrow isothermal flat plate will be discussed.

As before, the heat transfer from a narrow isothermal plate embedded in a plane adiabatic surface with the adiabatic surface being in the same plane as the heated plate will be considered. Attention will be given to the case where, in general, the plate is inclined at an angle to the vertical. The situation here considered is therefore as shown in Fig. 3.1. Results for a relatively wide range of Rayleigh numbers and dimensionless plate widths for both positive and negative inclination angles, i.e., both with the heated plate facing upwards and with the heated plate facing downwards as shown in Fig. 3.2 will be presented here. Attention has again been restricted to results for a Prandtl number of 0.7, this being approximately the value existing in the applications that originally motivated much of the work in this area.

As discussed in Chap. 1, there have been a number of studies of natural convective heat transfer from wide inclined plates. The results obtained in these studies have mainly been for a relatively narrow range of the governing parameters and until quite recently there still existed a need for a broader range of results

P. H. Oosthuizen and A. Y. Kalendar, *Natural Convective Heat Transfer*   31
from Narrow Plates, SpringerBriefs in Thermal Engineering and Applied Science,
DOI: 10.1007/978-1-4614-5158-7_3, © The Author(s) 2013

Fig. 3.1 Flow situation
considered

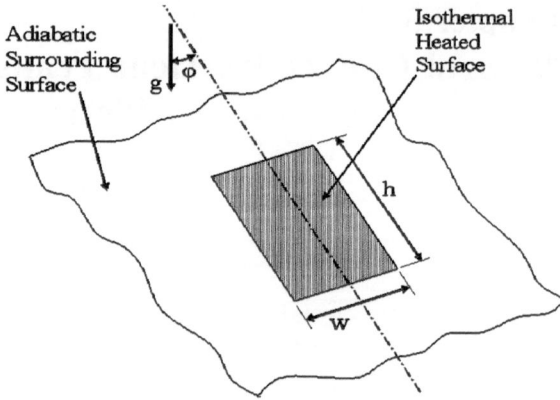

Fig. 3.2 Definition of
positive and negative angles of
inclination to the vertical, φ

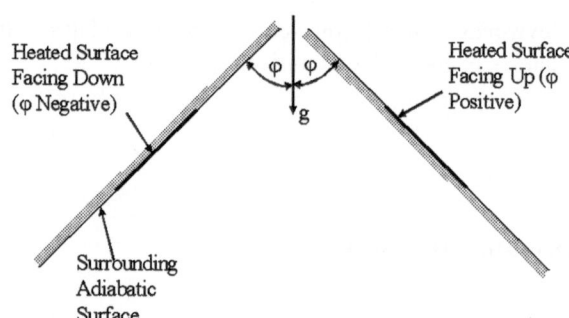

including results for narrow plates that can be used as the basis for predicting heat transfer rates in practical situations involving these types of flow. In the work being considered here, therefore, the natural convective heat transfer rates from narrow inclined and vertical isothermal plates for a relatively wide range of Rayleigh numbers and dimensionless plate widths were numerically determined. The discussion given in this chapter is largely based on the studies described in Kalendar and Oosthuizen [1, 2].

3.2 Solution Procedure

The same numerical procedure as that used in dealing with heat transfer from a vertical plate that was discussed in Chap. 2 has been adopted in the present work. The flow has been assumed to be steady and laminar. Fluid properties have been assumed constant except for the density change with temperature which gives rise to the buoyancy forces, this having been treated by using the Boussinesq approach. It has also been assumed that the flow is symmetric about the vertical center-plane of the plate. The solution has been obtained by numerically solving the full three-

Fig. 3.3 Solution domain

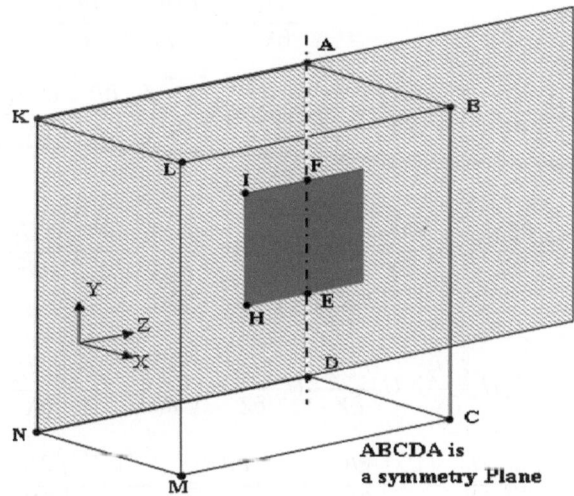

ABCDA is
a symmetry Plane

dimensional form of the governing equations, these equations being written in terms of dimensionless variables using the height, h, of the heated plate as the length scale and the overall temperature difference $(T_H - T_F)$ as the temperature scale, T_F being the fluid temperature far from the plate. Defining, as before, the following reference velocity:

$$u_r = \frac{\alpha}{h}\sqrt{Ra\,Pr} \tag{3.1}$$

where Pr is the Prandtl number and Ra is the Rayleigh number based on h, i.e.,:

$$Ra = \frac{\beta g(T_H - T_F)h^3}{\nu\alpha} \tag{3.2}$$

The following dimensionless variables have then again been defined:

$$X = \frac{x}{h},\ Y = \frac{y}{h},\ Z = \frac{z}{h},\ U_X = \frac{u_x}{u_r},$$
$$U_Y = \frac{u_y}{u_r},\ U_Z = \frac{u_z}{u_r},\ P = \frac{(p - p_F)h}{\mu u_r},\ \theta = \frac{T - T_F}{T_H - T_F} \tag{3.3}$$

where θ is the dimensionless temperature, T_H is the temperature of the hot wall and, as noted previously, T_F is the gas temperature far from the plate. The x-coordinate is measured in the horizontal direction normal to the plate, the y-coordinate is measures in the vertically upward direction and the z-coordinate is measured in the horizontal direction in the plane of the plate (see Fig. 3.3).

In terms of these dimensionless variables, the governing equations are if φ is the angle of the plate to the vertical:

$$\frac{\partial U_X}{\partial X} + \frac{\partial U_Y}{\partial Y} + \frac{\partial U_Z}{\partial Z} = 0 \tag{3.4}$$

$$U_X\frac{\partial U_X}{\partial X} + U_Y\frac{\partial U_X}{\partial Y} + U_Z\frac{\partial U_X}{\partial Z} = \sqrt{\frac{Pr}{Ra}}\left(-\frac{\partial P}{\partial X} + \frac{\partial^2 U_X}{\partial X^2} + \frac{\partial^2 U_X}{\partial Y^2} + \frac{\partial^2 U_X}{\partial Z^2}\right) + \theta\sin\varphi \tag{3.5}$$

$$U_X\frac{\partial U_Y}{\partial X} + U_Y\frac{\partial U_Y}{\partial X} + U_Z\frac{\partial U_Y}{\partial X} = \sqrt{\frac{Pr}{Ra}}\left(-\frac{\partial P}{\partial Y} + \frac{\partial^2 U_Y}{\partial X^2} + \frac{\partial^2 U_Y}{\partial Y^2} + \frac{\partial^2 U_Y}{\partial Z^2}\right) + \theta\cos\varphi \tag{3.6}$$

$$U_X\frac{\partial U_Z}{\partial X} + U_Y\frac{\partial U_Z}{\partial Y} + U_Z\frac{\partial U_Z}{\partial Z} = \sqrt{\frac{Pr}{Ra}}\left(-\frac{\partial P}{\partial Z} + \frac{\partial^2 U_Z}{\partial X^2} + \frac{\partial^2 U_Z}{\partial Y^2} + \frac{\partial^2 U_Z}{\partial Z^2}\right) \tag{3.7}$$

$$U_X\frac{\partial \theta}{\partial X} + U_Y\frac{\partial \theta}{\partial Y} + U_Z\frac{\partial \theta}{\partial Z} = \frac{1}{\sqrt{RaPr}}\left(\frac{\partial^2 \theta}{\partial X^2} + \frac{\partial^2 \theta}{\partial Y^2} + \frac{\partial^2 \theta}{\partial Z^2}\right) \tag{3.8}$$

Because the flow has been assumed to be symmetric about the vertical center-line of the plate the domain used in obtaining the solution is as shown in Fig. 3.3. Considering the surfaces shown in Fig. 3.3, the assumed boundary conditions on the solution in terms of the dimensionless variables are, since flow symmetry is being assumed:

$$\text{FEHI:} \quad U_X = 0, \; U_Y = 0, \; U_Z = 0, \; \theta = 1$$

$$\text{DNKA except for FEHI:} \quad U_X = 0, \; U_Y = 0, \; U_Z = 0, \; \frac{\partial \theta}{\partial X} = 0$$

$$\text{BCML:} \quad U_Y = 0, \; U_Z = 0, \; \theta = 0$$

$$\text{KLMN:} \quad U_X = 0, \; U_Y = 0, \; \theta = 0$$

$$\text{DCMN:} \quad U_X = 0, \; U_Z = 0, \; \theta = 0$$

$$\text{ABCDEF:} \quad U_Z = 0, \; \frac{\partial U_Y}{\partial Z} = 0, \; \frac{\partial U_X}{\partial Z} = 0, \; \frac{\partial \theta}{\partial Z} = 0$$

$$\text{ABLKA:} \quad P = 0$$

The mean and local heat transfer rate from the heated plate have been expressed in terms of the following mean and local Nusselt numbers:

$$Nu = \frac{\bar{q}'h}{k(T_H - T_F)}, \quad Nu_y = \frac{q'y}{k(T_H - T_F)} \tag{3.9}$$

where $\bar{q}'$ and q' are the mean and local heat transfer rate from the heated plate per unit area, respectively.

The dimensionless governing equations subject to the boundary conditions discussed above have been numerically solved using the commercial finite volume method based code, FLUENT©. To ensure the accuracy of the results presented, extensive grid and convergence criterion independence testing was undertaken.

This indicated that the heat transfer results presented here are to within 1 % independent of the number of grid points and of the convergence criterion used. The effect of the positioning of the outer surfaces of the solution domain (i.e., surfaces BCML, KLMN, DCMN, and ABLK shown in Fig. 3.3) from the heated surface was also examined and the positions used in obtaining the results discussed here were chosen to ensure that the heat transfer results were independent of this positioning to within 1 %.

An indication of the adequacy of the numerical model is also given by comparing the results obtained here for the case of natural convection over a vertical plate with existing experimental correlation equations that are available for wide vertical flat plates in the laminar flow region. Such a comparison is essentially made in the next section. A consideration of the results given there indicates that there is good agreement between the present numerical results with established experimental correlation equations.

3.3 Results

The solution has the following parameters:

1. The Rayleigh number, Ra, based on the plate height, h, and the overall temperature difference between the plate temperature and the fluid temperature
2. The dimensionless plate width, $W = w/h$
3. The Prandtl number, Pr
4. The angle of inclination of the plate from the vertical, φ

Results have only been obtained for $Pr = 0.7$. Results were then obtained for Ra values between 10^3 and 10^7, W values between 0.3 and 1.2, and inclination angles, φ, of between $-45°$ and $+45°$.

Figure 3.4 shows a comparison of the variation of the local Nusselt number with Rayleigh number for a wide plate along the y-direction at the center of the plate obtained in the present numerical study for $\varphi = 0°$ to $\varphi = +45°$ with existing correlations given by Hassan and Mohamed[3], Al-Arabi and Sakr [4], and similarity solution discussed by Oosthuizen and Naylor [5] for vertical and inclined flat plate facing up from $\varphi = 0°$ to $\varphi = +45°$. These results show good agreement with these existing results, the uncertainty in these existing results being up to 10 % for the angles of inclination considered.

The variations of the mean Nusselt number with Rayleigh number for various values of the dimensionless plate width, W, for inclination angles of $0°$, $+45°$, and $-45°$ are shown in Figs. 3.5, 3.6, and 3.7, respectively. The empirical correlation shown in these figures is the standard Churchill and Chu [6] correlation equation for natural convection from a vertical wide flat plate modified to apply to an inclined plate by replacing the Rayleigh number, Ra, in this equation by $Ra \cos \varphi$. This correlation then gives:

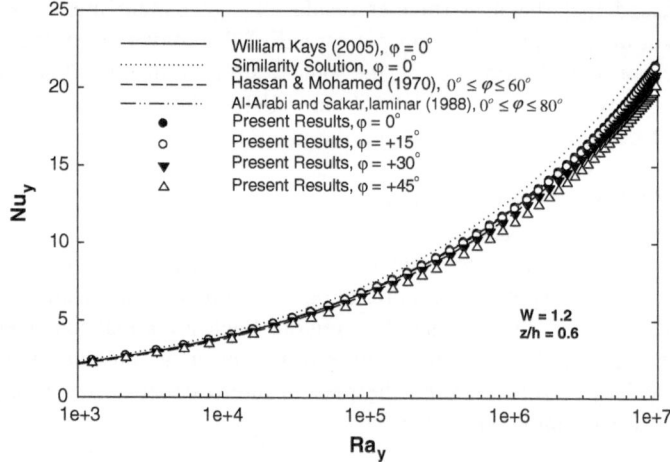

Fig. 3.4 Variations of numerical and experimental correlations result of local Nusselt number with Rayleigh number for $W = 1.2$ along the vertical center-line for a wide vertical and inclined flat plate facing up

$$Nu_0 = 0.68 + \left\{\frac{0.67}{\left[1 + (0.492/\operatorname{Pr})^{9/16}\right]^{4/9}}\right\}(Ra\cos\varphi)^{0.25} \qquad (3.10)$$

which becomes for $Pr = 0.7$:

$$Nu_0 = 0.68 + 0.513\,(Ra\cos\varphi)^{0.25} \qquad (3.11)$$

It will be seen from Fig. 3.5 that for $\varphi = 0°$, i.e., for a vertical plate, particularly at the lower Rayleigh numbers considered, the mean Nusselt number tends to increase with decreasing W. As mentioned before, the increase in the Nusselt number with decreasing W arises from the fact that there is an induced inflow toward the plate from the sides and this causes the heat transfer rate to be higher near the vertical sides of the plate than in the center region of the plate, i.e., edge effects become important. At the higher values of Ra, the edge effect becomes less important and consequently as the width of the plate, W, increases the results are in agreement with those given by the correlation equation for a wide plate.

From Fig. 3.6, it will be seen that for the upward facing heated plate at all values of Ra considered the Nusselt number is higher than that given by the correlation equation, the difference increasing with decreasing W. The difference between the effect of W on the Nusselt number for the upward facing plate and the vertical plate basically arises because with the inclined plate there is a component of the buoyancy force normal to the plate surface which gives rise to pressure changes across the flow over the plate, the pressure at the surface for the case of the upward facing being lower than in the fluid outside of the flow. This pressure

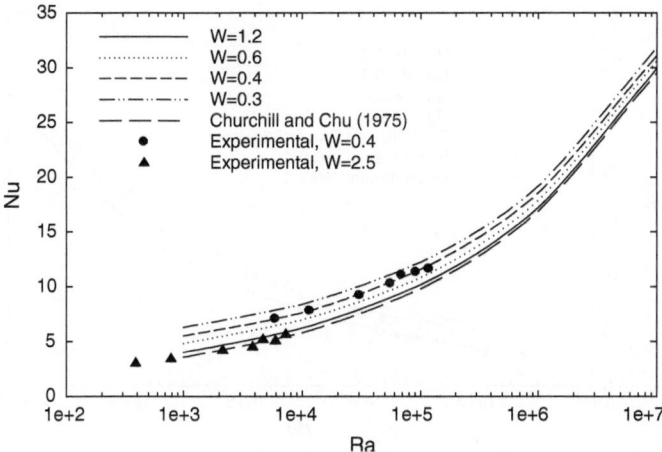

Fig. 3.5 Variation of mean Nusselt number with Rayleigh Number for various values of W for $\varphi = 0°$. Experimental results are discussed in Chap. 5

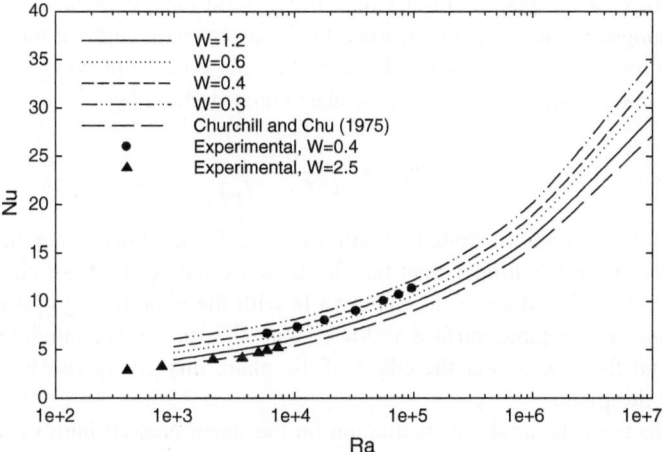

Fig. 3.6 Variation of mean Nusselt number with Rayleigh number for various values of W for $\varphi = +45°$. Experimental results are discussed in Chap. 5

difference causes the flow near the edges of the plate to be different from the flow over a vertical plate.

From Fig. 3.7, which shows results for the case of a downward facing plate, it will be seen that the variation of Nu with Ra is closer to that existing with a vertical plate but the effect of W on the results is again to be higher at larger values

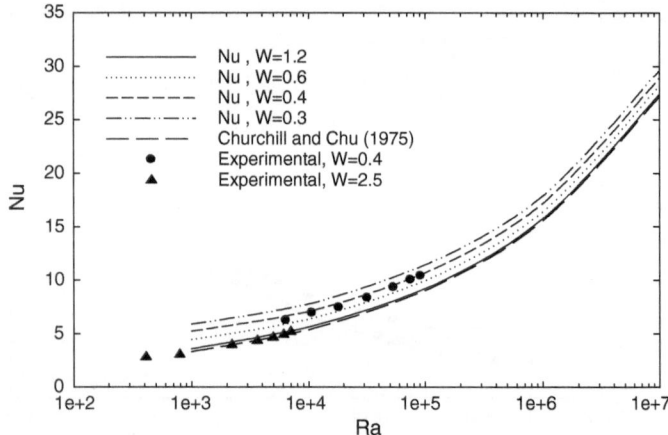

Fig. 3.7 Variation of mean Nusselt number with Rayleigh number for various values of W for $\varphi = -45°$. Experimental results are discussed in Chap. 5

of Ra than that for the vertical plate. With the downward facing plate, the pressure at the surface of the plate is higher than in the fluid outside of the flow.

The changes in the three-dimensional flow patterns with angle of inclination are illustrated by the local Nusselt number distributions over the heated plate shown in Figs. 3.8 and 3.9, the local Nusselt number being defined by:

$$Nu_L = \frac{q' h}{k(T_H - T_F)} \tag{3.12}$$

Figures 3.8 and 3.9 illustrate that with the plate facing down, when the pressure at the plate surface is higher than that in the surrounding fluid, an outward fluid flow exists near the edges of the plate while with the plate facing upwards, when the pressure at the plate surface is lower than that in the surrounding fluid, an inward fluid flow exist near the edges of the plate, this giving rise to a complex flow over the plate.

The effects of the angle of inclination on the mean Nusselt number are further illustrated by the results given in Figs. 3.10, 3.11, and 3.12 which show the variations of $Nu/(Ra \cos \varphi)^{0.25}$ with Ra for various values of the dimensionless plate width, W, for inclination angles of 0°, +45°, and −45°, respectively. These results show how, as a result of the edge effects, in all cases $Nu/(Ra \cos \varphi)^{0.25}$ increases with decreasing Ra and decreasing W. The results also show that the edge effects are different for an inclined plate than for a vertical plate and that the edge effect is different for the plate facing down than it is for the plate facing up.

Figure 3.13 shows the variation of the mean Nusselt number with angle of inclination for $Ra = 10^4$ for four values of W. It will be seen from this figure that in all cases Nu is smaller for a negative angle of inclination than it is for the corresponding positive angle of inclination.

Fig. 3.8 Variation of local Nusselt number distribution for $Ra = 10^6$ and $W = 0.3$ for inclination angles φ of $-45°$ and $+45°$

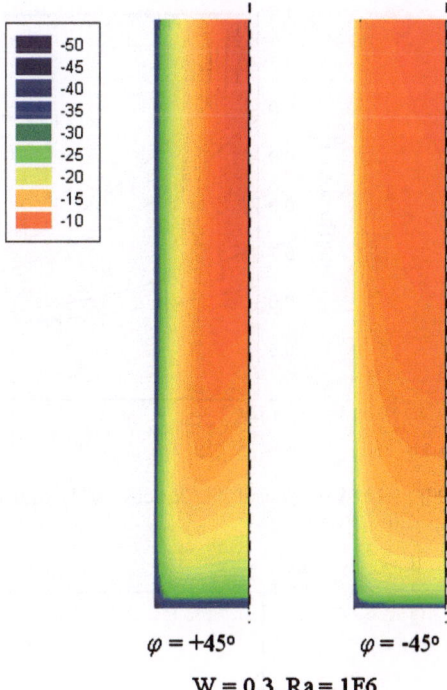

$$\varphi = +45° \qquad \varphi = -45°$$

$$W = 0.3, Ra = 1E6$$

Fig. 3.9 Variation of local Nusselt number distribution for $Ra = 10^6$ and $W = 0.6$ for inclination angles φ of $-45°$ and $+45°$

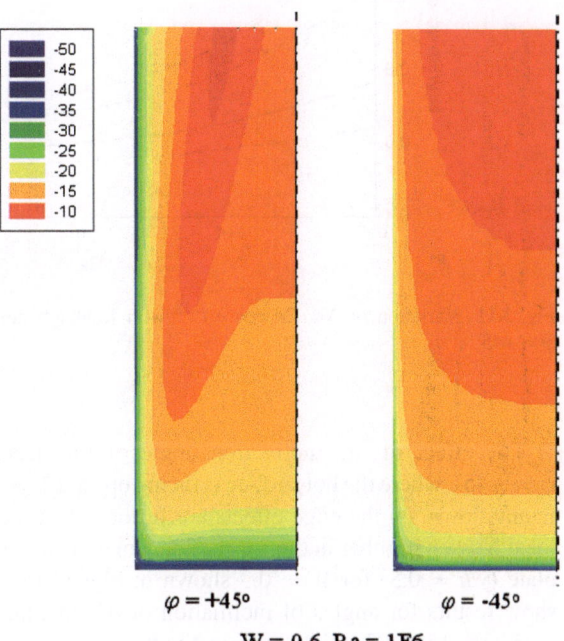

$$\varphi = +45° \qquad \varphi = -45°$$

$$W = 0.6, Ra = 1E6$$

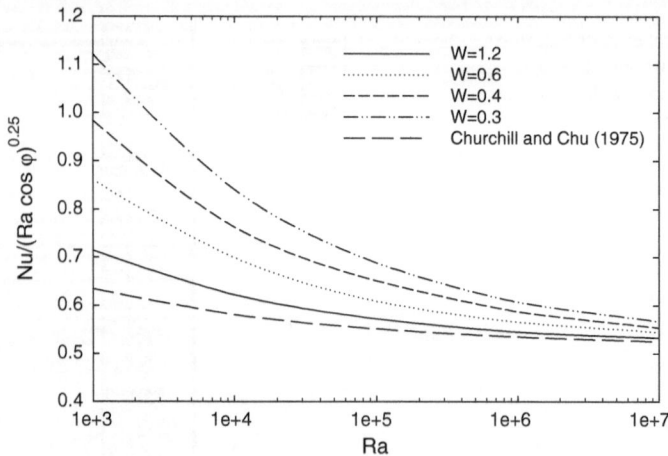

Fig. 3.10 Variation of $Nu/(Ra \cos \varphi)^{0.25}$ with Rayleigh number for various values of W for $\varphi = 0°$

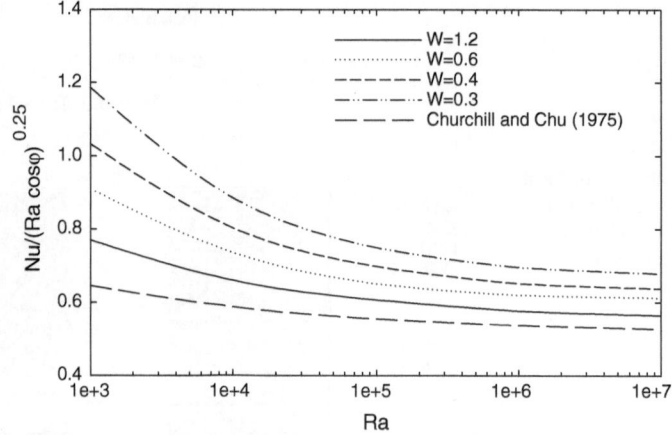

Fig. 3.11 Variation of $Nu/(Ra \cos \varphi)^{0.25}$ with Rayleigh number for various values of W for $\varphi = +45°$

The effect of the inclination angle of the heated plate for angles between $\varphi = +45°$, where the hot surface is facing up, and $\varphi = -45°$, where the hot surface is facing down, on the edge effects are further illustrated by the typical variations of local Nusselt number in the z-direction along the horizontal center-line of the heated plate ($y/h = 0.5$) for $W = 0.3$ shown in Figs. 3.14, 3.15, and 3.16. These figures show results for angles of inclination of $0°$, $45°$, and $-45°$, $0°$, $-15°$, $-30°$, and $-45°$; $0°$, $15°$, $30°$ and $45°$, respectively.

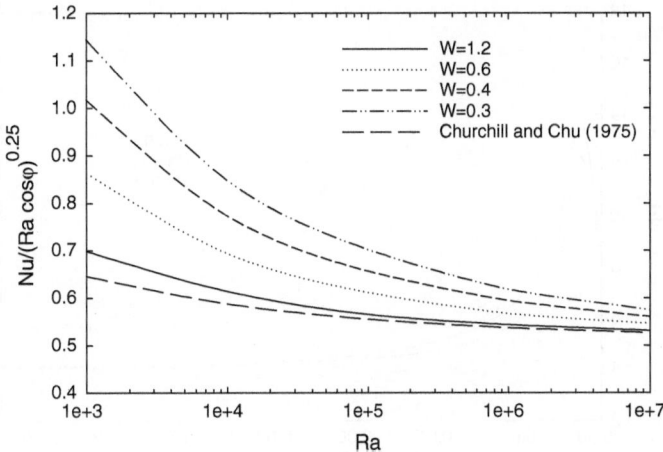

Fig. 3.12 Variation of $Nu/(Ra \cos \varphi)^{0.25}$ with Rayleigh number for various values of W for $\varphi = -45°$

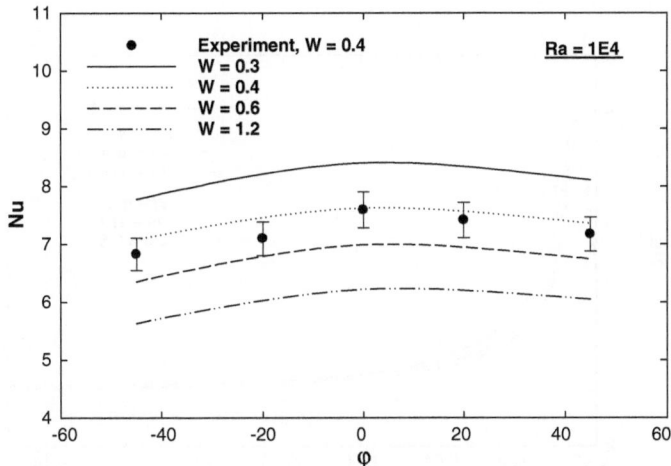

Fig. 3.13 Numerical variation of Nu with angle of inclination, φ, for $Ra = 1E4$ and for various values of W. Experimental results are discussed in Chap. 5

The results given in Fig. 3.14 show that the local Nusselt number when the plate is inclined facing up is higher than for a vertical plate and for an inclined plate facing down near the edge of the heated plate when $z/h < 0.04$. Although the buoyancy force parallel to the plate for the case of the inclined plate is lower than the buoyancy force for the vertical plate case, the local Nusselt number for the inclined plate facing up is higher than that for the vertical and the inclined facing down plate near the edge of the heated plate. When the heated plate is facing up

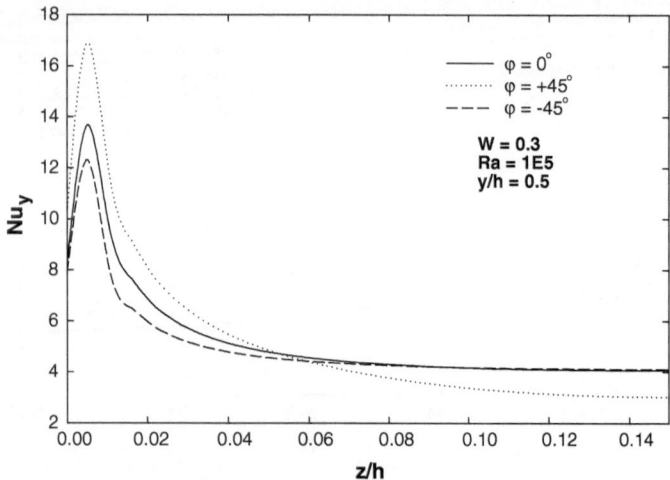

Fig. 3.14 Variation of local Nusselt number with dimensionless distance along the horizontal center-line between side edge and vertical center-line ($y/h = 0.5$) for inclination angle of 0°, −45° and +45° when $W = 0.3$ and $Ra = 1E5$

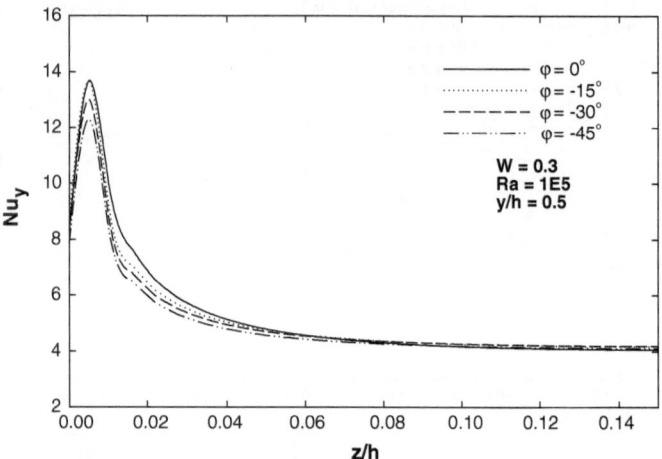

Fig. 3.15 Variation of local Nusselt number with dimensionless distance along the horizontal center-line between side edge and vertical center-line ($y/h = 0.5$) for inclination angle of 0°, −15°, −30°, and −45° when $Ra = 1E5$ and $W = 0.3$

the edge effect covers more of the plate surface area than when the heated plate is facing down. This is because when the plate is inclined at an angle to the vertical where the hot surface is facing up a buoyancy component exists in the direction away from the surface which causes the pressure at the heated plate surface to be lower than that in the surrounding fluid which results in a stronger inward fluid

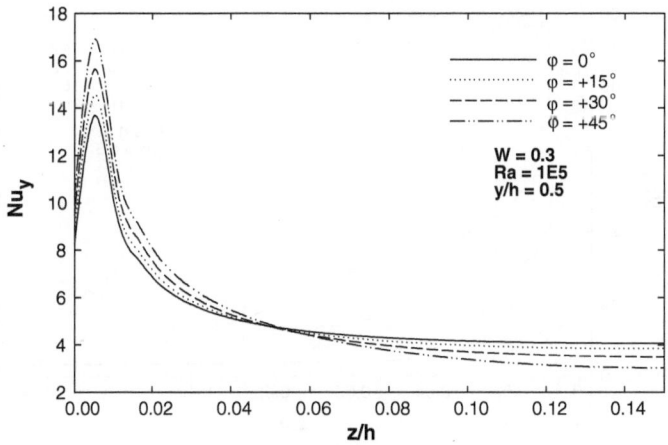

Fig. 3.16 Variation of local Nusselt number with dimensionless distance along the horizontal center-line between side edge and vertical center-line ($y/h = 0.5$) for inclination angle of 0°, +15°, +30°, and +45° when $Ra = 1E5$ and $W = 0.3$

flow which gives rise to a higher heat flux near the plate edge than exists with a vertical plate and an inclined plate facing down.

Figure 3.15 which shows variation of local Nusselt number with dimensionless distance z/h for inclination angles of 0°, −15°, −30°, and −45°, again indicates that the local Nusselt number has a higher value near the edge of the heated plate and that its value decreases as z/h increases. The results also show that the greater the negative inclination angle (heated plate facing down) the lower are the values of the local Nusselt number. Figure 3.16, which shows variations of local Nusselt number with dimensionless distance z/h for inclination angles of 0°, +15°, +30°, and +45°, indicates that as the angle of inclination increases where the hot surface is facing up the local heat transfer increases near the edge of the plate when $z/h < 0.04$. The results also show that as the angle of inclination increases the point at which the minimum local heat transfer occurs moves toward $z/h = 0.15$, the center of the heated plate, because as the inclination angle increases the pressure at the heated surface increases which increases the induced flow from the edge toward the center of the heated plate.

Figures 3.17 and 3.18 show the variations of local Nusselt number, Nu_y, with z/h for $W = 0.3$ and 0.4, for the case of $y/h = 0.5$ and $\varphi = +45°$ for different Rayleigh numbers, Ra. These results show that the position of maximum local Nusselt number is independent of the dimensionless plate width and Rayleigh number, while the position of minimum local Nusselt number is dependent on the dimensionless plate width and that the minimum and maximum values of the local Nusselt number increase as the Rayleigh number increases. The minimums that occur in the local Nusselt number distributions occur at the points where the flow separates from the surface, forming a plume in the upward direction, as discussed by Komori et al. [7] and Kimura et al. [8]. The low local heat transfer that occurs

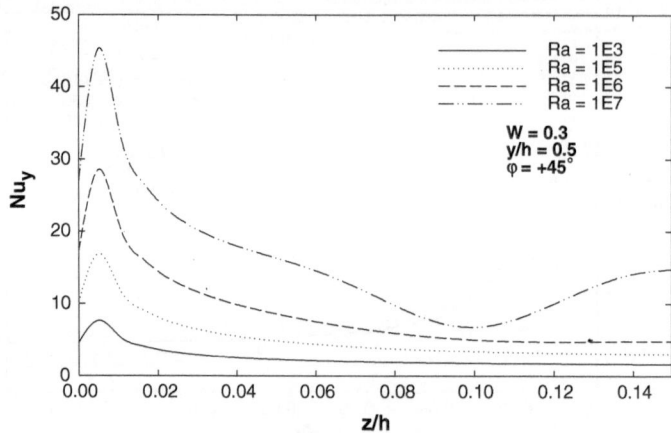

Fig. 3.17 Variation of local Nusselt Number with dimensionless distance along the horizontal center-line between side edge and vertical center-line ($y/h = 0.5$) for inclination angle of +45° and $W = 0.3$ for different Ra

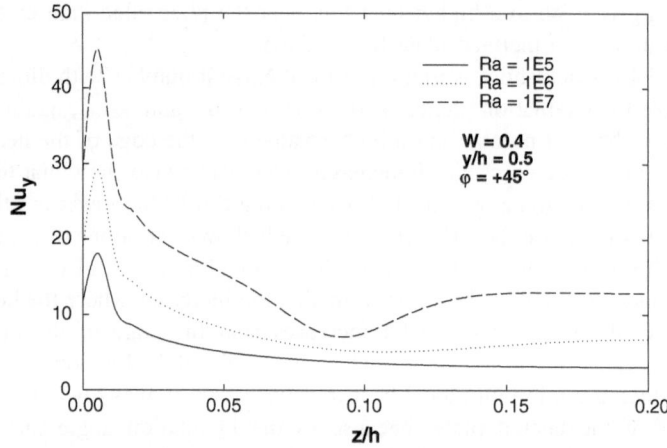

Fig. 3.18 Variation of local Nusselt number with dimensionless distance along the horizontal center-line between side edge and vertical center-line ($y/h = 0.5$) for inclination angle of +45° and $W = 0.4$ for different Ra

in the region of plume formation is illustrated by the local Nusselt number distributions shown in Fig. 3.19.

Now, the correlation equation for the case of a narrow vertical plate (i.e., $\varphi = 0°$) with a uniform surface temperature for a Prandtl number of 0.7 has the form:

$$\frac{Nu_0}{Ra^{0.25}} = \text{constant} + \text{function}\,(W, Ra) \tag{3.13}$$

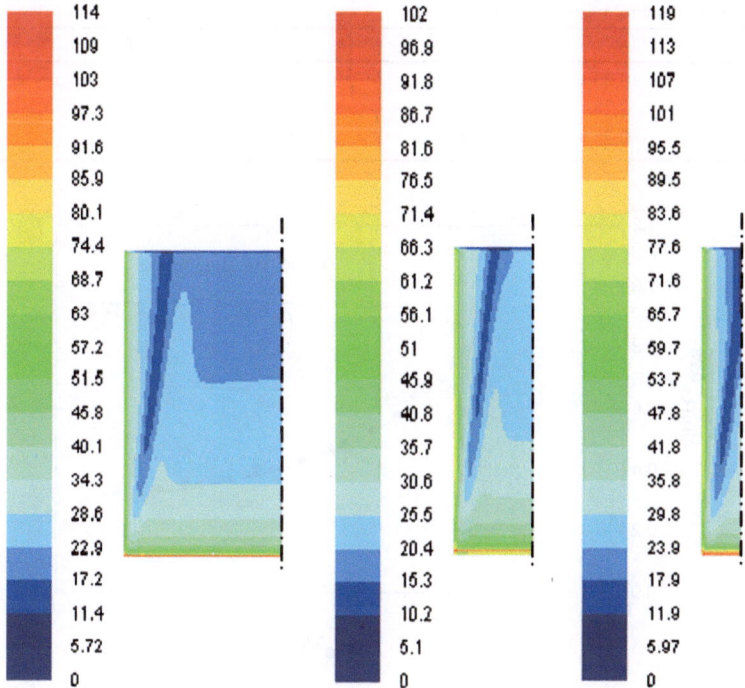

114	102	119
109	96.9	113
103	91.8	107
97.3	86.7	101
91.6	81.6	95.5
85.9	76.5	89.5
80.1	71.4	83.6
74.4	66.3	77.6
68.7	61.2	71.6
63	56.1	65.7
57.2	51	59.7
51.5	45.9	53.7
45.8	40.8	47.8
40.1	35.7	41.8
34.3	30.6	35.8
28.6	25.5	29.8
22.9	20.4	23.9
17.2	15.3	17.9
11.4	10.2	11.9
5.72	5.1	5.97
0	0	0

Fig. 3.19 Variation of local Nusselt number based on h over the heated plate for $Ra = 1E7$ and $\varphi = +45°$ for $W = 1.2$ (*left*), $W = 0.6$ (*center*) and $W = 0.3$ (*right*)

As discussed in Chap. 2, the numerical results for narrow vertical plates can be correlated using the following equation which is based on the form indicated in Eq. (3.13):

$$\frac{Nu_0}{Ra^{0.25}} = 0.51 + \frac{7.1}{(wRa^{0.5})}1.25 \tag{3.14}$$

Using an equation of this form, it has been found that the present numerical results for the case of vertical isothermal heated plate, $\varphi = 0°$, and for inclined plates at different angles to the vertical, φ, for φ values between $-45°$ and $\varphi = +45°$, with different dimensionless widths can be approximately described by an equation of the form:

$$\frac{Nu_{\text{mviemp}}}{Ra^{(D+C\sin\varphi)}} = D + \frac{E}{(WRa^{0.5})^F} \tag{3.15}$$

where B, C, D, F, and E are constants whose values depend on the angle of inclination. Typical values of these constants shown in Table 3.1.

Table 3.1 Constants used in Eq. (3.15).

φ	B	C	D	E	F
−45°	0.25	0.03	0.32	1.9	0.53
0°	0.28	–	0.29	1.8	0.5
+45°	0.25	0.08	0.18	1.7	0.5

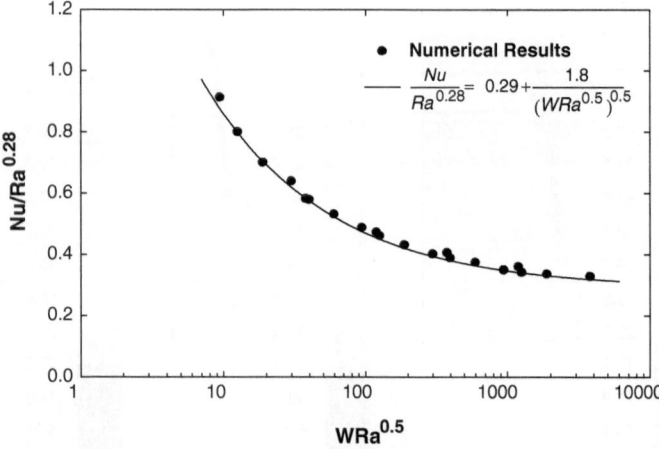

Fig. 3.20 Comparison of correlation equation for vertical flat plate, $\varphi = 0°$, with the numerical results

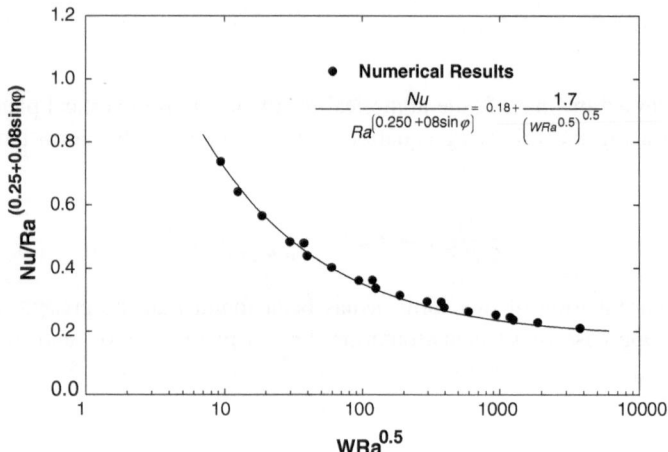

Fig. 3.21 Comparison of correlation equation for vertical flat plate, $\varphi = +45°$, with the numerical results

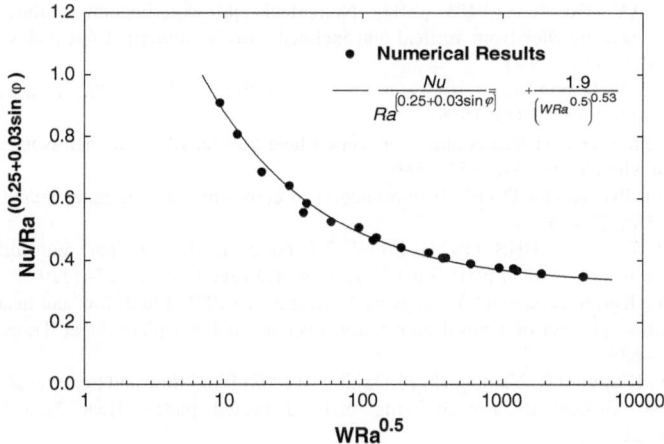

Fig. 3.22 Comparison of correlation equation for vertical flat plate, $\varphi = -45°$, with the numerical results

A comparison of the results given by Eq. (3.15) with the numerical results is shown in Figs. 3.20, 3.21, and 3.22. It will be seen that the equation describes the numerical results to an accuracy of better than 98 %.

3.4 Conclusions

Complex flows can arise over narrow inclined plates. However, for angles of inclination between $\varphi = -45°$ and $\varphi = +45°$ the mean heat transfer rates for this flow situation can be adequately described by the following equation:

$$\frac{Nu_{\text{mviemp}}}{Ra^{(B+C\sin\varphi)}} = D + \frac{E}{\left(WRa^{0.5}\right)^F}$$

where the values of the constants B, C, D, E, and F depend only on the angle of inclination. Some values of these constants are given in Table 3.1.

References

1. Kalendar AY, Oosthuizen PH (2008) Natural convective heat transfer from an inclined narrow isothermal flat plate. In: Jacksonville FL (ed) Proceedings of ASME national heat transfer conference, Paper HT2008-56190

2. Kalendar AY, Oosthuizen PH (2011) Numerical and experimental studies of natural convective heat transfer from vertical and inclined narrow isothermal flat plates. Heat Mass Transfer 47(9):1181–1195
3. Hassan KE, Mohamed SA (1970) Natural convection from isothermal flat surfaces. Int J Heat Mass Transfer 13(12):1873–1886
4. Al-Arabi M, Sakr B (1988) Natural convection heat transfer from inclined isothermal plates. Int J Heat Mass Transf 31(3):559–566
5. Oosthuizen PH, Naylor D (1999) Introduction to convective heat transfer analysis, 1st edn. McGraw-Hill, New York
6. Churchill SW, Chu HHS (1975) Correlating equations for laminar and turbulent free convection from a vertical plate. Int J Heat Mass Transfer 18(11):1323–1329
7. Komori K, Kito S, Nakamura T, Inaguma Y, Inagaki T (2001) Fluid flow and heat transfer in the transition process of natural convection over an inclined plate. Heat Trans Asian Res 30(8):648–659
8. Kimura F, Kitamura K, Yamaguchi M, Asami T (2003) Fluid flow and heat transfer of natural convection adjacent to upward-facing inclined heated plates. Heat Trans Asian Res 32(3):278–291

Chapter 4
Effect of Edge Conditions on Natural Convective Heat Transfer from Narrow Vertical Plates

Keywords Natural convection · Edge conditions · Narrow plates · Vertical plates · Isothermal · Uniform surface heat flux · Numerical

4.1 Introduction

In the preceding chapters, the discussion of heat transfer from narrow plates has been restricted to the situation where the heated plate is imbedded in a large plane adiabatic surface, the surfaces of the heated plane, and the adiabatic surface being in the same plane this flow situation being shown in Fig. 4.1. However, it is to be expected that the magnitude of the edge effects will depend, in general, on the boundary conditions existing near the edge of the plate. To initially investigate how important this effect is likely to be attention has been given to the natural convective heat transfer from a plate where there are only plane adiabatic surfaces above and below the heated plate and the vertical edges of the plate are thus directly exposed to the surrounding fluid, it being assumed that the plate and the adiabatic top and bottom surfaces are thin. This situation is shown in Fig. 4.2. In the cases considered, attention will first be given to the case where the plate is isothermal and then attention will be given to the case where there is a uniform heat flux over the plate surface.

Attention will also be given to natural convective heat transfer from a narrow isothermal vertical plate that protrudes from the surrounding adiabatic plane surface on which it is mounted, the case where the plate protrudes by a relatively small amount from the surrounding surface being considered.

The work discussed in this chapter is mainly based on the studies described by Oosthuizen and Paul [1–4].

P. H. Oosthuizen and A. Y. Kalendar, *Natural Convective Heat Transfer from Narrow Plates*, SpringerBriefs in Thermal Engineering and Applied Science, DOI: 10.1007/978-1-4614-5158-7_4, © The Author(s) 2013

Fig. 4.1 Situation
considered when there are
adiabatic side surfaces

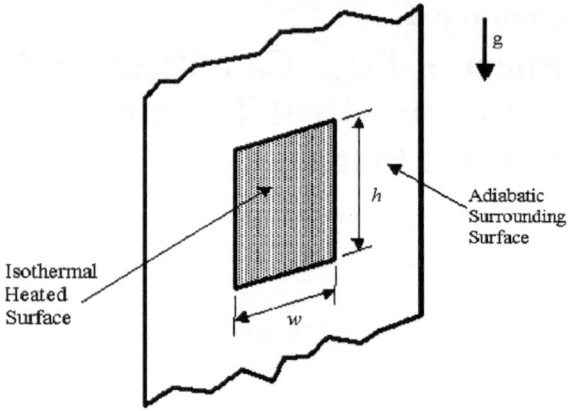

4.2 Vertical Isothermal Plates

Attention will first be given to a comparison of numerically derived results for the
two vertical plate situations shown in Figs. 4.1 and 4.2. The case where the plates
are isothermal will first be considered.

4.2.1 Solution Procedure

Basically, the same numerical procedure as described in earlier chapters has been
adopted. The flow has been assumed to be laminar and steady. It has been assumed
that the fluid properties are constant except for the density change with temperature which gives rise to the buoyancy forces, this having been treated by using the
Boussinesq approach. It has also been assumed that the flow is symmetric about
the vertical center-plane of the plate.

The solution domain used is shown in Fig. 4.3. Considering the surfaces shown
in Fig. 4.3, the assumed boundary conditions on the solution for the first edge
condition considered, i.e., the condition shown in Fig. 4.1, are presented as before,
in terms of the dimensionless variables introduced in earlier chapters and are, since
flow symmetry is being assumed:

Fig. 4.2 Situation
considered when there
are no side surfaces

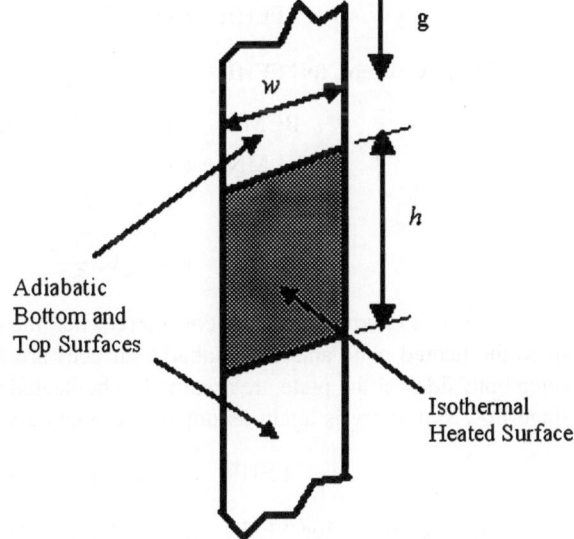

Fig. 4.3 Solution domain

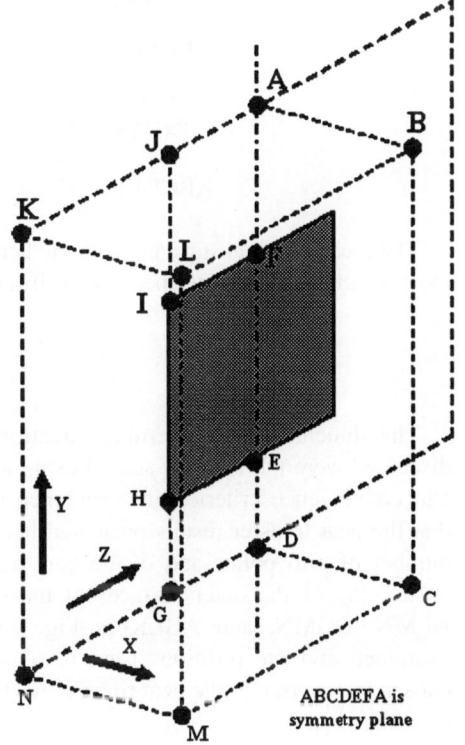

$$\text{FEHI:} \quad U_X = 0, U_Y = 0, U_Z = 0, \theta = 1$$

$$\text{DNKA except for FEHI:} \quad U_X = 0, U_Y = 0, U_Z = 0, \frac{\partial \theta}{\partial X} = 0$$

$$\text{BCML:} \quad U_Y = 0, U_Z = 0, \theta = 0$$

$$\text{KLMN:} \quad U_X = 0, U_Y = 0, \theta = 0$$

$$\text{DCMN:} \quad U_X = 0, U_Z = 0, \theta = 0$$

$$\text{ABCDEF:} \quad U_Z = 0, \frac{\partial U_Y}{\partial Z} = 0, \frac{\partial U_X}{\partial Z} = 0, \frac{\partial \theta}{\partial Z} = 0$$

For the second edge condition considered, i.e., the condition shown in Fig. 4.2, since the heated plate and the adiabatic surfaces are assumed to be very thin and since both sides of the plate are assumed to be heated to the same temperature and since flow symmetry is again assumed, the boundary conditions are:

$$\text{FEHI:} \quad U_X = 0, U_Y = 0, U_Z = 0, \theta = 1$$

$$\text{DGJA except for FEHI:} \quad U_X = 0, U_Y = 0, U_Z = 0, \frac{\partial \theta}{\partial X} = 0$$

$$\text{GNKJ:} \quad U_X = 0, \frac{\partial U_Y}{\partial X} = 0, \frac{\partial U_Z}{\partial X} = 0, \frac{\partial \theta}{\partial X} = 0$$

$$\text{BCML:} \quad U_Y = 0, U_Z = 0, \theta = 0$$

$$\text{KLMN:} \quad U_X = 0, U_Y = 0, \theta = 0$$

$$\text{DCMN:} \quad U_X = 0, U_Z = 0, \theta = 0$$

$$\text{ABCDEF:} \quad U_Z = 0, \frac{\partial U_Y}{\partial Z} = 0, \frac{\partial U_X}{\partial Z} = 0, \frac{\partial \theta}{\partial Z} = 0$$

The mean heat transfer rate from the heated plate has, as before, been expressed in terms of the following mean Nusselt number:

$$Nu = \frac{\bar{q}' h}{k(T_H - T_F)} \tag{4.1}$$

The dimensionless governing equations subject to the boundary conditions discussed were numerically solved using a commercial CFD code. Extensive grid and convergence criterion independence testing was undertaken. This indicated that the heat transfer results presented here are to within 1 % independent of the number of grid points and of the convergence criterion used. The effect of the positioning of the outer surfaces of the solution domain (i.e., surfaces MLBC, KLMN, DGMN, and ABLK in Fig. 4.3) from the heated surface was also examined and the positions used in obtaining the results discussed here were chosen to ensure that the heat transfer results were independent of this positioning to within 1 %.

Fig. 4.4 Typical variations
of mean Nusselt number with
W for various Rayleigh
number values for the two
edge conditions being
considered

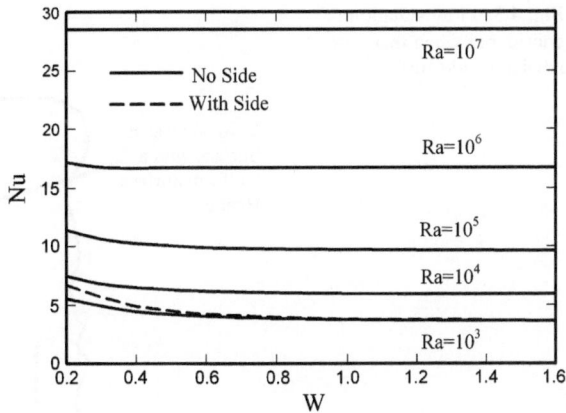

4.2.2 Results

The solution has the following parameters:

1. The Rayleigh number, *Ra*, based on the plate height, *h*, and the overall
   temperature difference between the plate temperature and the fluid temperature
2. The dimensionless plate width, $W = w/h$
3. The Prandtl number, *Pr*
4. The conditions at the edge of the plate

As already mentioned, results have only been obtained for $Pr = 0.7$. *Ra* values
between 10^3 and 10^8 and *W* values between 0.2 and 1.6 have again been
considered. Typical variations of the mean Nusselt number with dimensionless
plate width for various Rayleigh numbers for the two edge conditions that are
being considered are shown in Fig. 4.4. It will be seen from these results that,
particularly at the lower Rayleigh numbers, the mean Nusselt number tends to
increase with decreasing *W*. As mentioned before, the increase in the Nusselt
number with decreasing *W* arises from the fact that there is an induced inflow
toward the plate from the sides and this causes the heat transfer rate to be higher
near the vertical sides of the plate than it is in the center region of the plate.

It will also be noted from Fig. 4.4 that, except at the lowest Rayleigh number
considered, i.e., 10^3, at small values of *W* the results for the two edge conditions
considered are the same.

4.2.3 Concluding Remarks

The plate edge conditions, for the two edge conditions considered in this section,
only affect the heat transfer rate from the plate when the Rayleigh number and
dimensionless plate width are very small.

Fig. 4.5 Flow situation
considered when there are
adiabatic side surfaces

Vertical Heated
Surface with a
Uniform Surface
Heat Flux

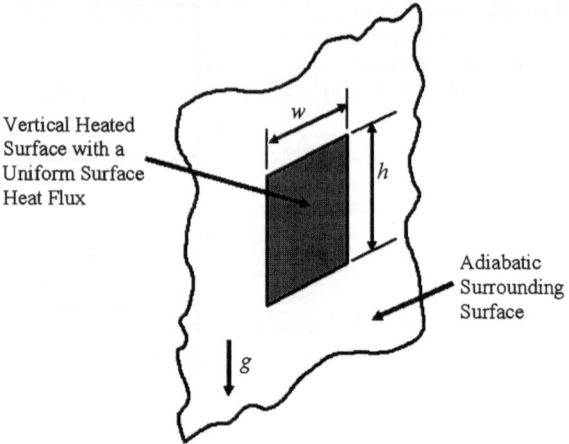

Adiabatic
Surrounding
Surface

4.3 Vertical Plates with a Uniform Surface Heat Flux

The case of heat transfer from an isothermal plate was considered in the previous
section. Now, it is possible that the thermal conditions at the plate surface will
affect the nature of the edge effects. For this reason, attention will next be given to
the effect of plate edge conditions on natural convective heat transfer rates from
narrow vertical plates which have a uniform surface heat flux over the surface.
Attention has again been restricted to results for a Prandtl number of 0.7, this being
approximately the value existing in the applications that originally motivated the
interest in this problem.

The same two situations that were considered in dealing with heat transfer from
an isothermal plate will be considered here, i.e., the case where the heated plate is
imbedded in a large plane adiabatic surface, the surfaces of the heated plate, and
the adiabatic surface being in the same plane and the case where there are only
plane adiabatic surfaces above and below the heated plate and the vertical edges of
the plate are directly exposed to the surrounding fluid have been considered. These
two flow situations are shown in Figs. 4.5 and 4.6. The plate has been assumed to
be, in general, narrow, i.e., the width of the plate, w, is assumed to be of the same
order of magnitude as the vertical height of the plate, H. In the second situation
considered it is assumed that the plate and the adiabatic top and bottom surfaces
are thin.

4.3.1 Solution Procedure

The same assumptions as used in dealing with the isothermal plate case have been
adopted here, i.e., the flow has been assumed to be laminar and it has been
assumed that the fluid properties are constant except for the density change with

Fig. 4.6 Flow situation
considered when there are
no side surfaces

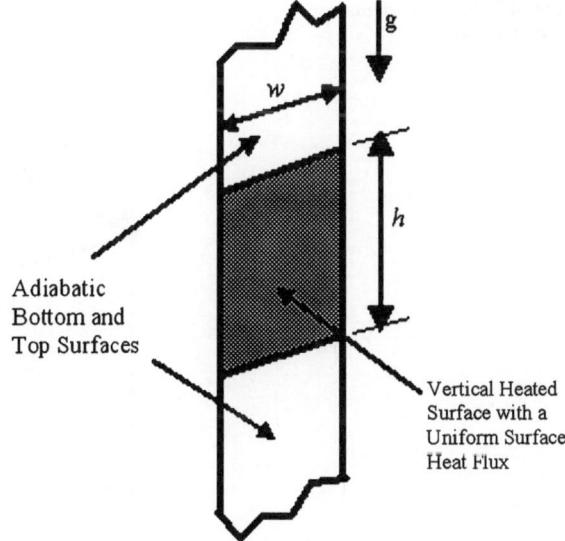

temperature which gives rise to the buoyancy forces, this having been treated by
using the Boussinesq approach. It has also been assumed that the flow is symmetric
about the vertical center-plane of the plate. The solution has been obtained by
numerically solving the full three-dimensional form of the governing equations,
these equations being written in terms of dimensionless variables using $q'H/k$ as
the temperature scale, q' being the uniform heat flux over the surface of the plate.
The heat flux Rayleigh number, Ra^* is then a parameter in this solution, this
Rayleigh number being defined by:

$$Ra^* = \frac{\beta g q' h^4}{k \nu \alpha} \tag{4.2}$$

Because the flow has been assumed to be symmetric about the vertical center-
line of the plate, the solution domain used in obtaining the solution is the same as
that used in the isothermal plate analysis, i.e., is as shown in Fig. 4.7. ABCDEFA
is the center-plane about which the flow is assumed to be symmetric.

Considering the surfaces shown in Fig. 4.7, the assumed boundary conditions
on the solution for the first edge condition considered are, in terms of the
dimensionless variables used in the analysis, since flow symmetry is being
assumed:

Fig. 4.7 Solution domain
ABCDIJLM

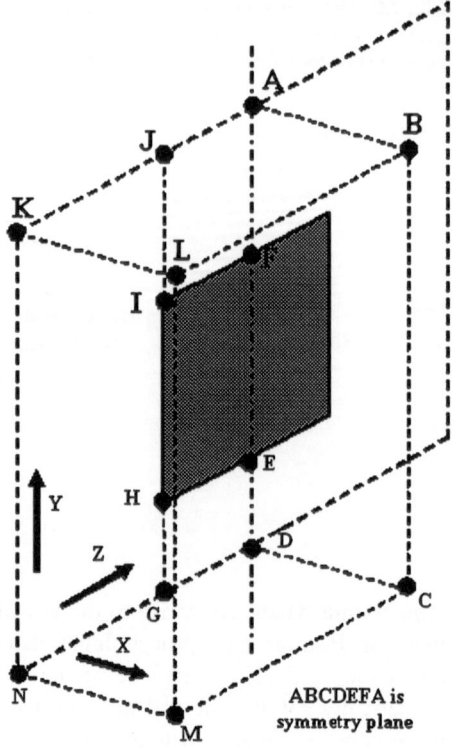

ABCDEFA is
symmetry plane

$$\text{FEHI:} \quad U_X = 0, U_Y = 0, U_Z = 0, \frac{\partial \theta}{\partial X} = -1$$

$$\text{DNKA except for FEHI:} \quad U_X = 0, U_Y = 0, U_Z = 0, \frac{\partial \theta}{\partial X} = 0$$

$$\text{BCML:} \quad U_Y = 0, U_Z = 0, \theta = 0$$

$$\text{KLMN:} \quad U_X = 0, U_Y = 0, \theta = 0$$

$$\text{DCMN:} \quad U_X = 0, U_Z = 0, \theta = 0$$

$$\text{ABCDEF:} \quad U_Z = 0, \frac{\partial U_Y}{\partial Z} = 0, \frac{\partial U_X}{\partial Z} = 0, \frac{\partial \theta}{\partial Z} = 0$$

For the second edge condition considered, since the heated plate and the adiabatic surfaces are assumed to be very thin and since conditions on both sides of the plate are assumed to the same, the boundary conditions in this case are assumed to be:

Fig. 4.8 Variation of mean Nusselt number with dimensionless plate width for various values of the heat flux Rayleigh number for the two edge conditions being considered

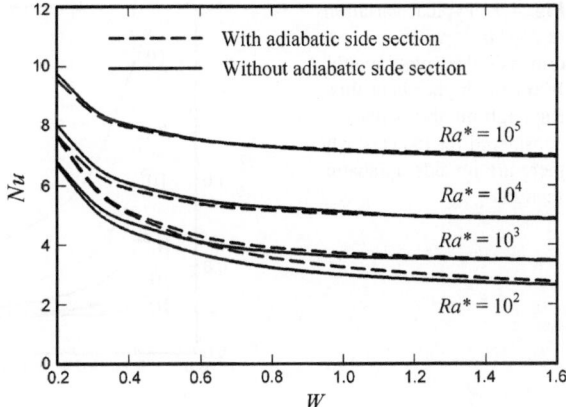

$$\text{FEHI:}\quad U_X = 0, U_Y = 0, U_Z = 0, \frac{\partial\theta}{\partial X} = -1$$

$$\text{DGJA except for FEHI:}\quad U_X = 0, U_Y = 0, U_Z = 0, \frac{\partial\theta}{\partial X} = 0$$

$$\text{GNKJ:}\quad U_X = 0, \frac{\partial U_Y}{\partial X} = 0, \frac{\partial U_Z}{\partial X} = 0, \frac{\partial\theta}{\partial X} = 0$$

$$\text{BCML:}\quad U_Y = 0, U_Z = 0, \theta = 0$$

$$\text{KLMN:}\quad U_X = 0, U_Y = 0, \theta = 0$$

$$\text{DCMN:}\quad U_X = 0, U_Z = 0, \theta = 0$$

$$\text{ABCDEF:}\quad U_Z = 0, \frac{\partial U_Y}{\partial Z} = 0, \frac{\partial U_X}{\partial Z} = 0, \frac{\partial\theta}{\partial Z} = 0$$

The mean heat transfer rate from the heated plate has been expressed in terms of the following mean Nusselt number:

$$Nu = \frac{q'h}{k(T_{Hm} - T_F)} = \frac{1}{\theta_{Hm}} \tag{4.3}$$

where T_{Hm} and θ_{Hm} are the mean temperature and the mean dimensionless temperature of the heated surface, respectively.

The dimensionless governing equations subject to the boundary conditions discussed above were numerically solved using a commercial CFD solver. Extensive grid and convergence criterion independence testing was undertaken. This indicated that the heat transfer results presented here are again to within 1 % independent of the number of grid points and of the convergence criterion used. The effect of the positioning of the outer surfaces of the solution domain (i.e., surfaces MLBC, KLMN, DCMN, and ABLK in Fig. 4.7) from the heated surface was also examined and the positions used in obtaining the results discussed here were chosen to ensure that the heat transfer results were independent of this positioning to within 1 %.

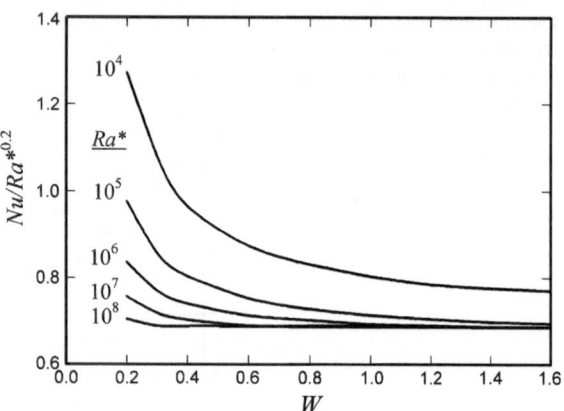

Fig. 4.9 Typical variations of $Nu/Ra^{*0.2}$ with dimensionless plate width W for the higher heat flux Rayleigh number values considered for the case where there are no side adiabatic sections

4.3.2 Results

The solution has the following parameters:

1. The heat flux Rayleigh number, Ra^*, based on the plate height, h, and the surface heat flux, q'
2. The dimensionless plate width, $W = w/h$
3. The Prandtl number, Pr
4. The conditions at the edge of the plate

As already mentioned, results have only been obtained for $Pr = 0.7$. Ra^* values between 10^3 and 10^8 and W values between 0.2 and 1.6 have been considered.

Typical variations of the mean Nusselt number for the plate with dimensionless plate width for various heat flux Rayleigh numbers for the two side conditions being considered are shown in Fig. 4.8. It will be seen from these results that in all cases as noted in the earlier chapter, the mean Nusselt number tends to increase with decreasing W at a given heat flux Rayleigh number.

It will further be noted that at the lower heat flux Rayleigh numbers considered, the Nusselt numbers for the case where there is no side section are lower than those for the case where there is a side section whereas at the higher heat flux Rayleigh numbers the opposite is true. The rest of the discussion will focus on results for the higher heat flux Rayleigh numbers, i.e., for Ra^* values of 10^4 and higher.

The effect of dimensionless plate width on the Nusselt number is further illustrated by the results shown in Figs. 4.9 and 4.10 for these higher heat flux Rayleigh numbers. In these figures, it has been noted that with two-dimensional flow over a wide vertical plate that has a uniform heat flux over its surface:

$$\frac{Nu}{Ra^{*0.2}} = \text{function}(Pr) \qquad (4.4)$$

Fig. 4.10 Typical variations of $Nu/Ra^{*0.2}$ with dimensionless plate width W for the higher heat flux Rayleigh number values considered for the case where there are side adiabatic sections

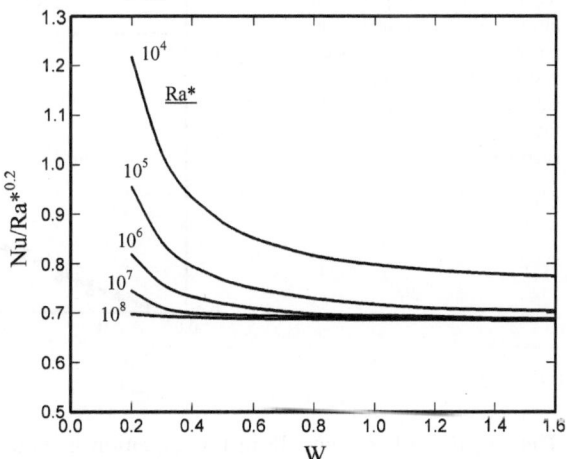

Figures 4.9 and 4.10, therefore, show the variations of $Nu/Ra^{*0.2}$ with the dimensionless plate width, W, for various values of Ra^* for the case where there is no adiabatic side section and for the case where there is an adiabatic side section. It will be seen that at the larger values of W and Ra^* the quantity $Nu/Ra^{*0.2}$ does tend to a constant value of approximately 0.68 which is the value given by the similarity solution for two-dimensional flow over a wide plate with a uniform heat flux at the surface for a Prandtl number of 0.7. It will also be seen from Figs. 4.9 and 4.10 that at the lowest values of W and Ra^* considered the value of $Nu/Ra^{*0.2}$ is more than 80 % above the two-dimensional flow value. However, at the highest value of Ra^* considered the two-dimensional value applies at W values down to about 0.3.

In order to obtain an approximate equation for the Nusselt number for narrow plates, it has been assumed that for wide plates, i.e., for situations in which the edge effects are negligible, Nu is given for $Pr = 0.7$, as mentioned before, by:

$$Nu = 0.68\, Ra^{*0.2} \qquad (4.5)$$

As discussed above, this equation gives results that are in good agreement with the present numerical results at the larger values of W and Ra^* considered. Because of this assumed form of the variation of Nu with Ra^* when edge effects are negligible, it has again been assumed that the Nusselt number for a plate of arbitrary dimensionless width W is given by an equation of the form:

$$\frac{Nu}{Ra^{*0.2}} = 0.68 + \text{function}(Ra^*, W) \qquad (4.6)$$

Fig. 4.11 Comparison of results given by Eq. 4.9 with numerical results for the no adiabatic side section case and with the results given by Eq. 4.8 for the case where there are adiabatic side sections

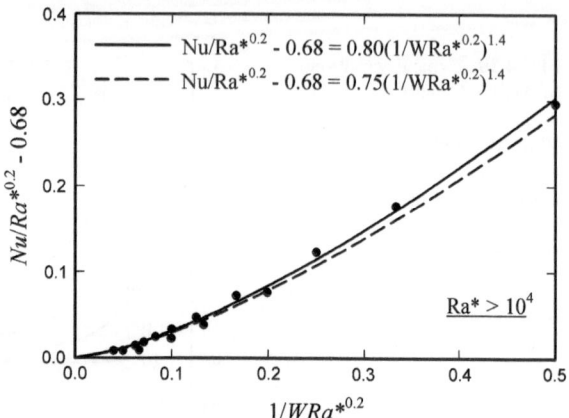

The function of Ra^* and W in this equation is assumed to depend on $1/WRa^{*0.2}$. The above equation has, therefore, been assumed to have the form:

$$\frac{Nu}{Ra^{*0.2}} = 0.68 + \text{function}\left(\frac{1}{WRa^{*0.2}}\right) \tag{4.7}$$

The form of the function was determined from the numerical results which indicated that the Nusselt number results obtained numerically could be approximately described for the case where there are side adiabatic sections by:

$$\frac{Nu}{Ra^{*0.2}} = 0.68 + \frac{0.75}{(WRa^{*0.2})^{1.4}} \tag{4.8}$$

and for the case where there are no side adiabatic sections by:

$$\frac{Nu}{Ra^{*0.2}} = 0.68 + \frac{0.80}{(WRa^{*0.2})^{1.4}} \tag{4.9}$$

This equation is compared with some of the numerical results for the case where there is no side adiabatic section in Fig. 4.11. The results given by Eq. 4.8 for the case where there are adiabatic side sections are also shown in this figure.

If three-dimensional effects (i.e., edge effects) are assumed to be negligible if:

$$\frac{Nu/Ra^{*0.2} - Nu_0/Ra^{*0.2}}{Nu_0/Ra^{*0.2}} = \frac{Nu/Ra^{*0.2} - 0.68}{0.68} < 0.02 \tag{4.10}$$

i.e., if:

$$\frac{Nu}{Ra^{*0.2}} - 0.68 < 0.014 \tag{4.11}$$

which gives, by using Eq. 4.8, for the case where there are side adiabatic sections:

Fig. 4.12 Variation of W value above which edge effects are negligible with Ra^* for the two edge conditions considered

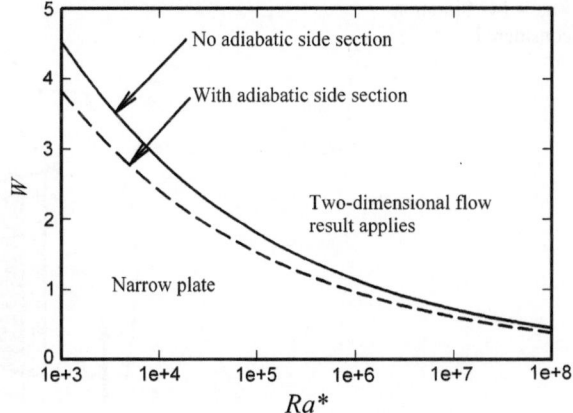

Fig. 4.13 Variation of ratio of Nusselt number for the no side section case to that for the case where there are side sections with $1/WRa^{*0.2}$ for Ra^* values of 10^4 and higher

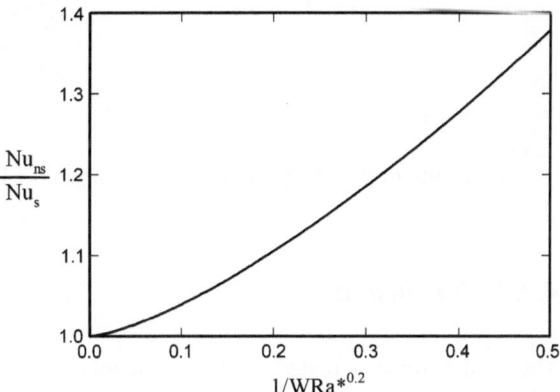

$$\frac{0.75}{\left(WRa^{*0.2}\right)^{1.4}} < 0.014 \quad \text{i.e.,} \quad WRa^{*0.2} > 15.2 \quad \text{i.e.,} \quad W > \frac{15.2}{Ra^{*0.2}} \quad (4.12)$$

while for the case where there are no side sections, Eq. 4.9 gives:

$$\frac{0.80}{\left(WRa^{*0.2}\right)^{1.4}} < 0.014 \quad \text{i.e.,} \quad WRa^{*0.2} > 18.0 \quad \text{i.e.,} \quad W > \frac{18.0}{Ra^{*0.2}} \quad (4.13)$$

Using these results, the variations of the values of W above which three-dimensional effects on Nu are negligible (to within approximately 2 %) with Ra^* can be determined for the two cases being considered and are shown in Fig. 4.12.

It will also be noted that using Eqs. 4.8 and 4.9 gives the following approximate relation for the ratio of the Nusselt number for the no side adiabatic section case to that for the with side adiabatic section case as:

$$\frac{Nu_{ns}}{Nu_s} = 1 + \frac{0.074}{\left(WRa^{*0.2}\right)^{1.4}} \quad (4.14)$$

Fig. 4.14 Situation
considered

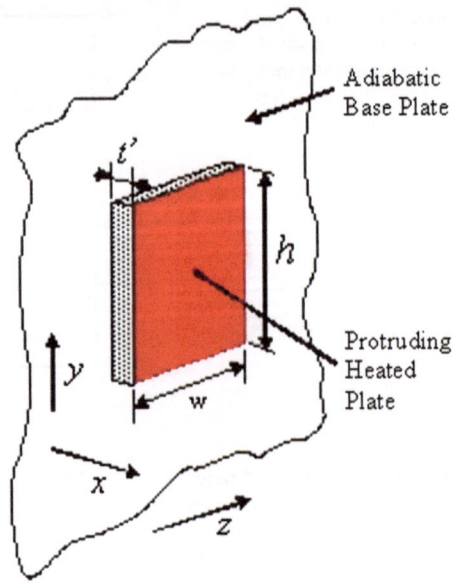

The results given by this equation are shown in Fig. 4.13.

4.3.3 Conclusions

The dimensionless plate width has again been shown to have a significant influence on the mean Nusselt number for natural convective heat transfer from a vertical flat plate with a uniform heat flux at the surface for both the case where there are side adiabatic sections and for the case where there are no adiabatic side sections. These edge effects increase with decreasing dimensionless plate width and decreasing Rayleigh number. The effect of the conditions at the sides of the plate when there is a uniform heat flux over the plate surface have been shown to be greater than when the plate is isothermal. Empirical equations for the mean heat transfer rate from narrow plates have been derived from the numerical results. These equations indicate that three-dimensional effects will be negligible for the case where there is a side section if:

$$W > \frac{15.2}{Ra^{*0.2}}$$

and for the case where there is no side section if:

$$W > \frac{18.0}{Ra^{*0.2}}$$

Fig. 4.15 Center-plane
section through plate
arrangement considered

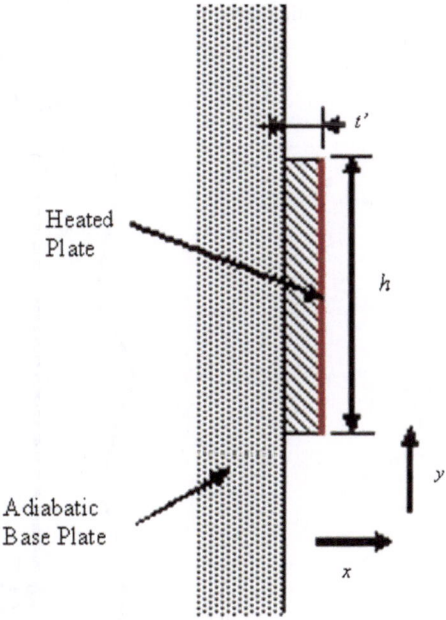

4.4   Protruding Vertical Isothermal Plates

The edge effects that arise in natural convective heat transfer from a narrow
isothermal vertical plate can also be dependent on whether the plate protrudes
from the surrounding adiabatic plane surface on which it is mounted or whether it
is recessed into this surrounding surface. To illustrate this effect attention will be
given to case where the plate protrudes by a relatively small amount from the large
surrounding plane adiabatic surface. The flow situation considered is thus as
shown in Figs. 4.14 and 4.15. Figure 4.14 shows the overall situation considered
while Fig. 4.15 shows a vertical cross-section through the system. The width of the
plate, w, is assumed to be of the same order of magnitude as the vertical height of
the plate, h, and the "height" that the plate is protrudes, t' has been assumed to be
relatively small compared to the plate height, h. It should be noted that Fig. 4.15 is
a schematic view of the situation considered and that the value of t' shown in
Fig. 4.15 is approximately 0.15 which is larger than the largest value of t' for
which results have here been obtained this being 0.1.

4.4.1 Solution Procedure

The same basic assumptions as used in the other numerical studies discussed in
this book have been adopted. The flow has been assumed to be laminar and it has

Fig. 4.16 Solution domain
used

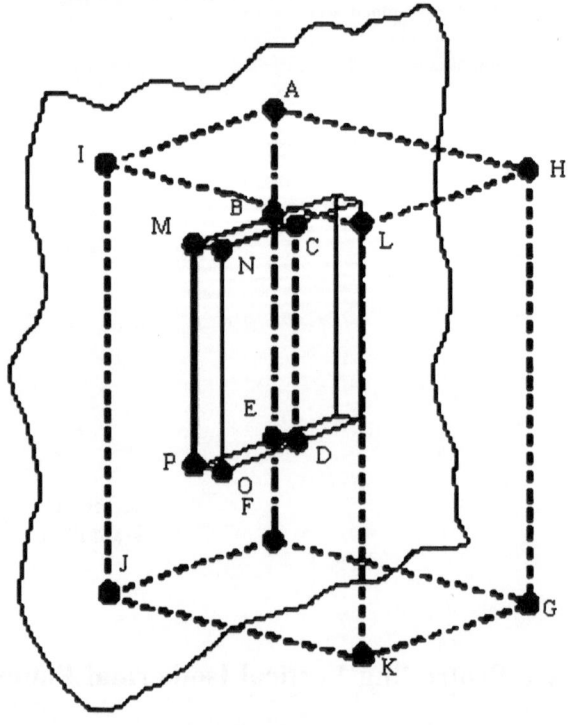

ABCDEFGHA is Symmetry Plane

been assumed that the fluid properties are constant except for the density change
with temperature which gives rise to the buoyancy forces, this having been treated
by using the Boussinesq approach. It has also been assumed that the flow is
symmetric about the vertical center-plane of the plate. The solution has been
obtained by numerically solving the full three-dimensional form of the governing
equations, these equations being written in terms of dimensionless variables using
the height, h, of the heated plate as the length scale and the overall temperature
difference $(T_H - T_F)$ as the temperature scale, T_F being the fluid temperature far
from the plate.

Because the flow has been assumed to be symmetric about the vertical center-
line of the plate the solution domain used in obtaining the solution is as shown in
Fig. 4.16.

The boundary conditions are basically that the velocity is zero on all solid
surfaces, that the dimensionless temperature on the heated plate is 1, that the
temperature gradients normal to all the remaining solid surfaces are zero, that the
dimensionless temperature is zero on surfaces FGKJF, HGKLH, and ILKJI shown
in Fig. 4.16 and that U_z and the gradients in the Z direction of all other dimensionless variables are zero on the plane of symmetry ABCDEFGHA, i.e.,:

$$\text{CDONC:} \quad U_X = 0, U_Y = 0, U_Z = 0, \theta = 1$$

$$\text{ABMPEFJIA:} \quad U_X = 0, U_Y = 0, U_Z = 0, \frac{\partial \theta}{\partial X} = 0$$

$$\text{NOPMN:} \quad U_X = 0, U_Y = 0, U_Z = 0, \frac{\partial \theta}{\partial Z} = 0$$

$$\text{EDOPE and BCNMB:} \quad U_X = 0, U_Y = 0, U_Z = 0, \frac{\partial \theta}{\partial Y} = 0$$

$$\text{HGKLH:} \quad U_Y = 0, U_Z = 0, \theta = 0$$

$$\text{LKJIL:} \quad U_X = 0, U_Y = 0, \theta = 0$$

$$\text{GKJFG:} \quad U_X = 0, U_Z = 0, \theta = 0$$

$$\text{ABCDEFGHA:} \quad U_Z = 0, \frac{\partial U_Y}{\partial Z} = 0, \frac{\partial U_X}{\partial Z} = 0, \frac{\partial \theta}{\partial Z} = 0$$

The mean heat transfer rate from the heated plate has again been expressed in terms of the following mean Nusselt number:

$$Nu = \frac{q'h}{k(T_H - T_F)} \tag{4.15}$$

The dimensionless governing equations subject to the boundary conditions discussed above have been numerically solved using the commercial finite-element solver, FIDAP. Extensive grid and convergence criterion independence testing was undertaken. This indicated that the heat transfer results presented here are to within 1 % independent of the number of grid points and of the convergence criterion used. The effect of the positioning of the outer surfaces of the solution domain (i.e., surfaces HGKLH, LKJIL, GKJFG, and AHLIA in Fig. 4.16) from the heated surface was also examined and the positions used in obtaining the results discussed here were chosen to ensure that the heat transfer results were independent of this positioning to within 1 %.

4.4.2 Results

The solution has the following parameters:

- The Rayleigh number, Ra
- The dimensionless plate width, W
- The dimensionless "height" to which the plate protrudes, $t = t'/h$
- The Prandtl number, Pr

Results have only been obtained for $Pr = 0.7$. Ra values between 10^3 and 10^7, W values between 0.2 and 1.2, and t values between 0 and 0.09 have been considered.

Fig. 4.17 Variation of mean
Nusselt number with W for
$t = 0.09$ for various Rayleigh
numbers

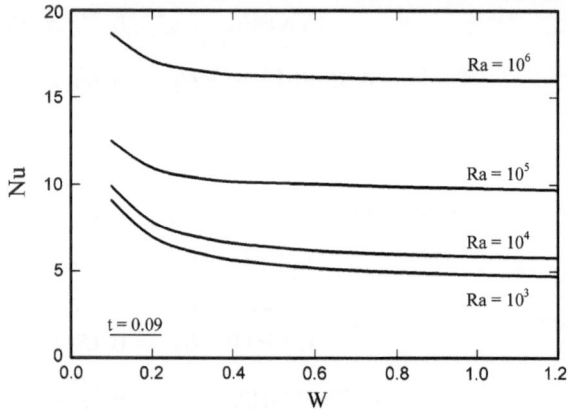

Fig. 4.18 Variation of mean
Nusselt number with W for
$t = 0.06$ for various Rayleigh
numbers

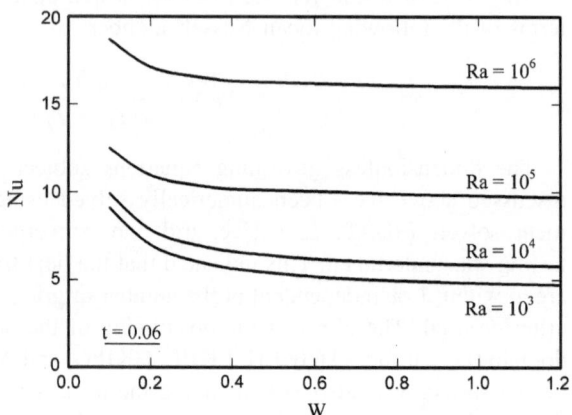

Fig. 4.19 Variation of mean
Nusselt number with W for
$t = 0.03$ for various Rayleigh
numbers

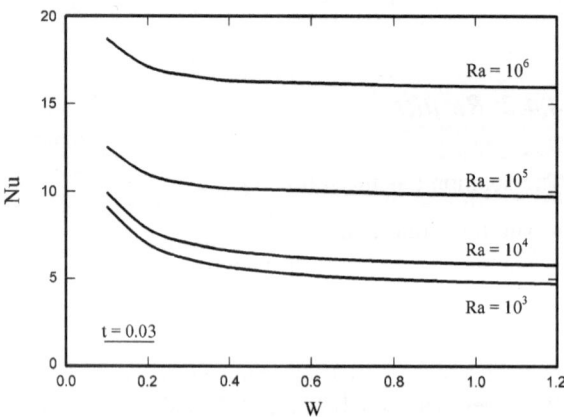

Fig. 4.20 Variation of mean Nusselt number with t for various values of W for $Ra = 10^6$

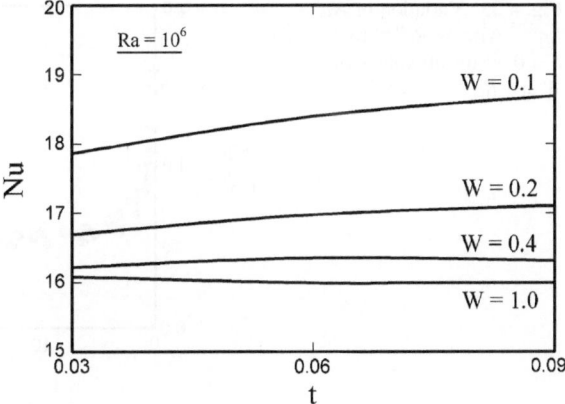

Fig. 4.21 Variation of mean Nusselt number with t for various values of W for $Ra = 10^4$

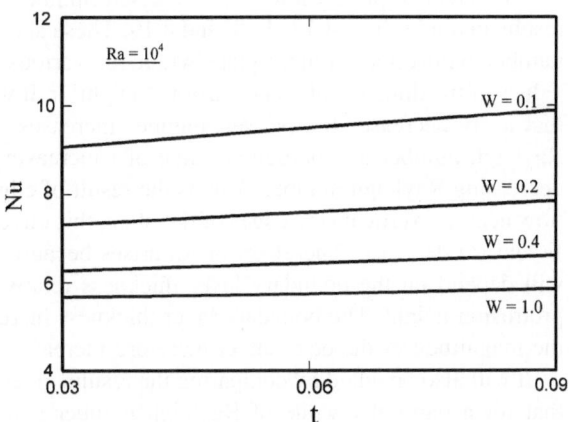

Fig. 4.22 Variation of mean Nusselt number with t for various values of W for $Ra = 10^3$

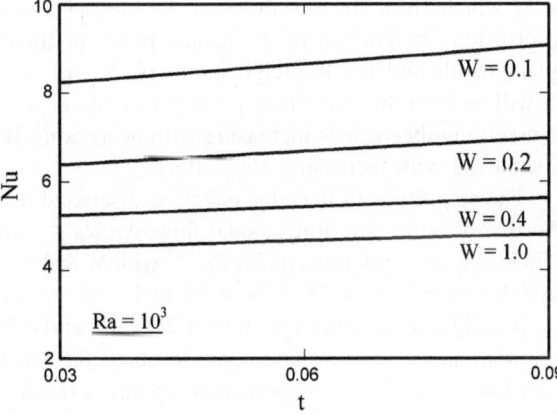

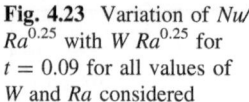

Fig. 4.23 Variation of $Nu/Ra^{0.25}$ with $W\,Ra^{0.25}$ for $t = 0.09$ for all values of W and Ra considered

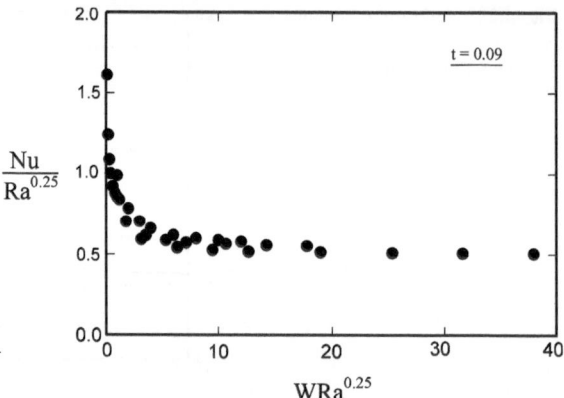

The effect of plate width on the Nusselt number variation is illustrated by the results given in Figs. 4.17, 4.18, and 4.19. These show the variation of the Nusselt number with dimensionless plate width for various Rayleigh numbers for three values of the dimensionless protrusion "height" t. It will be seen from these figures that as W decreases the Nusselt number increases, the increase for a particular Rayleigh number and particular value of t increasing in relative magnitude with decreasing Rayleigh number. This is the result of edge effects, i.e., of the induced flow near the vertical side edges of the plate, this effect increasing in importance as W and Ra decrease. The effect of Ra arises because the extent of the edge effect will depend on the boundary layer thickness relative to the plate width and the protrusion height. The boundary layer thickness increases with decreasing Ra and the magnitude of the edge effect therefore increases with decreasing Ra.

It will also be noted by comparing the results given in Figs. 4.17, 4.18, and 4.19 that for a particular value of Rayleigh number and a particular value of W the Nusselt number varies with the dimensionless protrusion height t. This effect of the protrusion of the plate is illustrated by the results given in Figs. 4.20, 4.21, and 4.22 which show the variations of the mean Nusselt number with dimensionless protrusion "height", t, of the heated plate for three values of the dimensionless plate width and for Rayleigh numbers, Ra, of 10^6, 10^4, and 10^3, respectively. It will be seen that the Nusselt number increases as t increases, the increase in the Nusselt number values increasing with decreasing W and with decreasing Ra, i.e., increasing with increasing edge effect.

The magnitude of the edge effect, as discussed before, will depend on $W\,Ra^{0.25}$. Furthermore, in two-dimensional flow, $Nu/Ra^{0.25}$ will be approximately constant. Therefore, the variations of $Nu/Ra^{0.25}$ with $W\,Ra^{0.25}$ for $t = 0.09$, 0.06, 0.03, and 0 are shown in Figs. 4.23, 4.24, 4.25, and 4.26, respectively.

It will be seen from Figs. 4.23, 4.24, 4.25, and 4.26 that in all cases $Nu/Ra^{0.25}$ is approximately constant and equal to about 0.51 at the larger values of $W\,Ra^{0.25}$ considered. As $W\,Ra^{0.25}$ decreases a point is reached at which $Nu/Ra^{0.25}$ starts to increase, i.e., edge effects begin to become significant. It will be seen from Figs. 4.23, 4.24, 4.25, and 4.26 that the value of $W\,Ra^{0.25}$ at which the increase in

Fig. 4.24 Variation of $Nu/Ra^{0.25}$ with $W\,Ra^{0.25}$ for $t = 0.06$ for all values of W and Ra considered

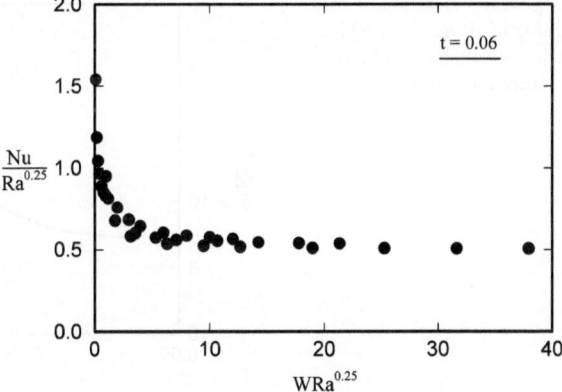

Fig. 4.25 Variation of $Nu/Ra^{0.25}$ with $W\,Ra^{0.25}$ for $t = 0.03$ for all values of W and Ra considered

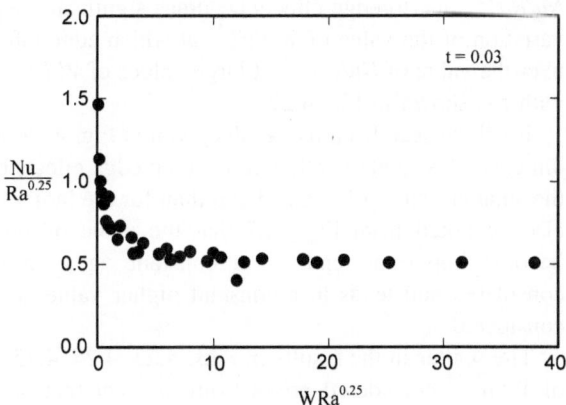

Fig. 4.26 Variation of $Nu/Ra^{0.25}$ with $W\,Ra^{0.25}$ for $t = 0$ for all values of W and Ra considered

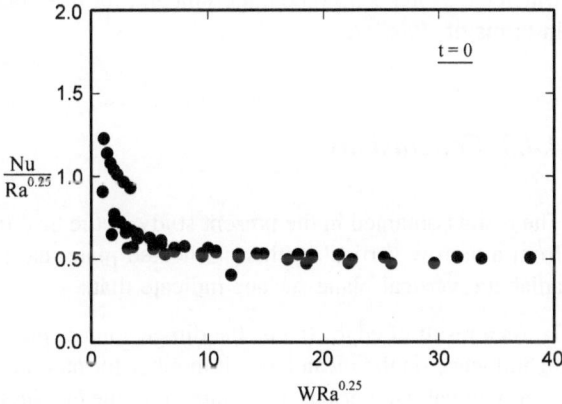

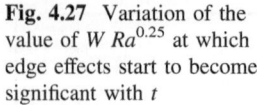

Fig. 4.27 Variation of the value of $W\,Ra^{0.25}$ at which edge effects start to become significant with t

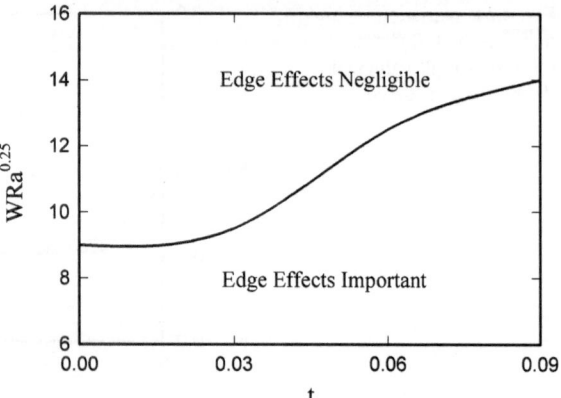

$Nu/Ra^{0.25}$ due to edge effects becomes significant depends on the value of t. The variation of the value of $W\,Ra^{0.25}$ at which edge effects, based on a 5 % increase over the value of $Nu/Ra^{0.25}$ at large values of $W\,Ra^{0.25}$, start to become significant with t is shown in Fig. 4.27.

It will be seen from the results given in Fig. 4.27 that at a given value of Ra the dimensionless plate width below which edge effects become important is lower at the smaller values of t considered than for the larger values of t considered. It will also be noted from Fig. 4.27 that the value of $WRa^{0.25}$ at which edge effects become important tends to a constant value at the low values of $WRa^{0.25}$ considered and tends to a constant higher value at the larger values of $WRa^{0.25}$ considered.

The scatter in the results in Figs. 4.23, 4.24, 4.25, and 4.26 at the lower values of $WRa^{0.25}$ considered arises from the fact that, as has been noted before, the results at these lower values of $WRa^{0.25}$ are better correlated in terms of $WRa^{0.5}$. However, it appears that in the situation here considered the conditions under which edge effects become important are, for most purposes, adequately correlated in terms of $WRa^{0.25}$.

4.4.3 Conclusions

The results obtained in the present study of the heat transfer by natural convection from a narrow vertical isothermal heated plate that protrudes from a surrounding adiabatic vertical plane surface indicate that:

1. As a result of edge effects, the dimensionless plate width can have a significant influence on the mean Nusselt number for natural convective heat transfer from a vertical isothermal flat plate, the mean Nusselt number increasing with decreasing dimensionless plate width at a given Rayleigh number.

2. The magnitude of the edge effect, i.e., the magnitude of the mean Nusselt number increase with decreasing dimensionless plate width, increases with decreasing Rayleigh number.
3. The dimensionless distance that the heated plate protrudes "above" the surrounding adiabatic surface has an influence on the magnitude of the edge effect and on the condition under which edge effects become important.

References

1. Oosthuizen PH, Paul JT (2007) Effect of edge conditions on natural convective heat transfer from a narrow vertical flat plate with a uniform surface heat flux. In: Proceedings of ASME international mechanical engineering congress and exposition, pp 397–404, Paper IMECE 2007-42712
2. Oosthuizen PH, Paul JT (2007) Natural convective heat transfer from a narrow vertical isothermal flat plate with different edge conditions. In: Proceedings of 15th annual computational fluid dynamics society of Canada conference (CFD), Toronto
3. Oosthuizen PH, Paul JT (2007) Natural convective heat transfer from a recessed narrow vertical flat plate with a uniform heat flux at the surface. In: Proceedings of 5th international heat transfer, fluid mechanics and thermodynamics conference (HEFAT 2007), Sun City, South Africa
4. Oosthuizen PH, Paul JT (2010) Natural convective heat transfer from a narrow vertical flat plate with a uniform surface heat flux and with different plate edge conditions. Frontiers Heat Mass Transf. doi:10.5098/hmt.v1.1.3006

Chapter 5
Experimental Results for Natural Convective Heat Transfer from Vertical and Inclined Narrow Plates

Keywords Natural convection · Experimental · Narrow flat plates · Inclined plates · Positive and negative inclination angles · Isothermal

5.1 Introduction

Various numerical studies of situations involving natural convective heat transfer from narrow plates are described in other chapters. However, in order to fully verify the adequacy of the numerical approach used in these studies an experimental study of natural convective heat transfer from vertical and inclined narrow isothermal flat plates in the laminar and transition flow regions was undertaken and is described in this chapter.

The heat transfer from a narrow vertical isothermal plate embedded in a plane adiabatic surface, with the adiabatic surface being in the same plane as the heated plate (see Fig. 5.1) and, in general, inclined at an angle to the vertical has been considered. Experimental results have been obtained for a range of Rayleigh numbers and for two dimensionless plate width-to-height ratios. Both positive and negative inclination angles, i.e., both the case where the heated plate is facing up and the case where the heated plate is facing down (see Fig. 5.2) have been considered. The experimental results have been obtained for the case where the plate is transferring heat to the surrounding air so that the experimental results, like the numerical results discussed in the other chapters, are essential for a Prandtl number of 0.7.

While there have been some past studies of the natural convective heat transfer rate from narrow vertical heated plates, (see discussion of past work given in Chap. 1) the results obtained in these studies have been for a relatively narrow range of the governing parameters and still there appeared to be a need for a broader range of results that could be used to adequately validate the present

P. H. Oosthuizen and A. Y. Kalendar, *Natural Convective Heat Transfer from Narrow Plates*, SpringerBriefs in Thermal Engineering and Applied Science, DOI: 10.1007/978-1-4614-5158-7_5, © The Author(s) 2013

73

Fig. 5.1 Flow situation
considered

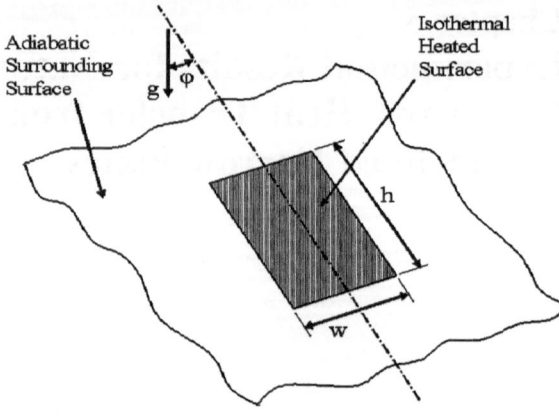

Fig. 5.2 Definition of
positive and negative angles
of inclination

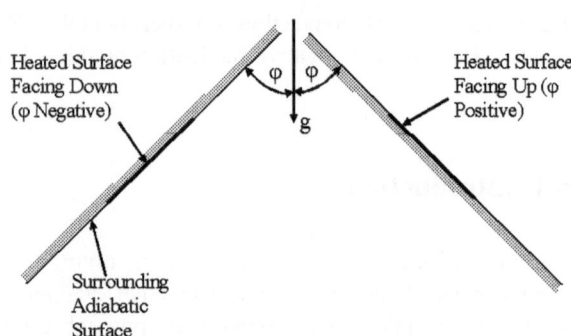

numerical results. The work discussed in this chapter is based on the study
described by Kalendar and Oosthuizen [1].

5.2 Experimental Apparatus and Procedure

To study the natural convection heat transfer from both vertical and inclined plates
to the vertical, a natural convection test chamber in which the model could be
mounted was designed and constructed. The volume of the test chamber was large
enough to ensure that the chamber walls did not interfere with the flow around the
test plate. The test chamber was constructed in such a way that it could be rotated
around a fixed horizontal axis, in this way allowing the inclination angle of the test
plate to the vertical to be changed. The test chamber was mounted in a larger
chamber that was designed to ensure that the external disturbances in the air in the
room surrounding the chamber and short-term temperature changes in this room
did not interfere with the flow near the test plate.

The test plates were basically made from aluminum with an electrical heater
attached to the rear surface of the plates. These plates with heater attached were

Fig. 5.3 Cross-section of
plane view of the test section
containing the aluminum
plate and the heater
embedded in acrylic
polycarbonate

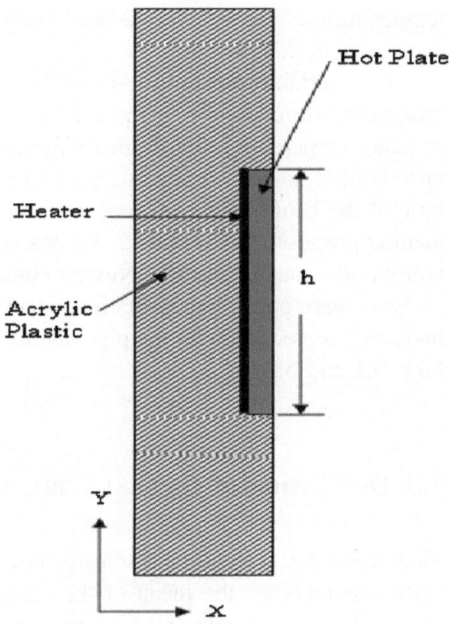

imbedded in a 20 cm by 24 cm acrylic polycarbonate layer that had a thickness of
25.4 mm. The basic layout is shown schematically in Fig. 5.3. Two plates were
used in obtaining the results described in this chapter. One plate had a height of
31.8 mm and a width of 12.7 mm while the other plate had a height of 12.7 mm
and a width of 31.8 mm, both aluminum plates having a thickness of 6 mm
thickness. This thickness was estimated to be adequate to ensure that the plate had
a uniform surface temperature during a test. The aspect ratio w/h of the two plates
used was thus 0.4 and 2.5. Very thin (0.23 mm thick) polyimide flexible 12.7 mm
by 31.8 mm micro heaters were attached to the back of the plates. The power
supplied to the heater was measured to an accuracy of ± 0.3 % of the measured
value. Electric power was supplied to the heater through a variable DC transformer
that was set to give an average plate temperature of the order required in a
particular test.

Temperatures at various locations within the heated plate were measured using
type-T thermocouples (0.25 mm in diameter). Four thermocouples were positioned
inside the shallow holes in the test plate just beneath the exposed surface on each
side of the aluminum test plate. The thermocouples were monitored using a data
acquisition system which was self-calibrating. This system was connected to a
computer system. The ambient air temperature was also monitored using a thermocouple mounted in the test chamber. The thermocouples with the data acquisition system were calibrated in a digital temperature controlled water bath.

In a given test once a steady-state condition had been reached, the average
mean temperature over the plate was measured. It was found that the maximum

temperature difference over the plate surface was in all cases less than 2.5 % of the average plate temperature.

The uncertainty in the experimentally determined values of the Nusselt number arises due to uncertainties in the temperature measurements, due to nonuniformities in plate temperature, due to uncertainties in the power measurement, and due to uncertainties in the corrections applied for conduction heat transfer through the base of the heated plate. An uncertainty analysis was performed by applying the method proposed by Moffat [2, 3]; this analysis indicating that the overall uncertainty in the mean measured Nusselt number was less than ±4 %.

Tests were performed with components mounted vertically, i.e., parallel to the buoyancy force, and with the plate inclined at an angle to the vertical as shown in Figs. 5.1 and 5.2.

5.3 Derivation of the Experimental Results

Since only the average heat transfer rate was being measured, the plate temperature was taken as the mean of the measured temperatures. Using the measured undisturbed test chamber air temperature, the mean film temperature could be then determined and the fluid properties of the air at this temperature is determined.

Now, the total rate of heat transfer from the plate by convection is given by

$$Q_{conv} = Q_{total} - Q_{loss} = Q_{heater} - Q_{loss} \qquad (5.1)$$

where

$$Q_{loss} = Q_{cond} + Q_{rad}, \ Q_{cond} = K(T_{W_{avg}} - T_F), \ Q_{rad} = \varepsilon \sigma A \left(T_{W_{avg}}^4 - T_F^4 \right) \quad (5.2)$$

Q_{cond} is the conductive heat transfer to the rear and side coverings of the plate and Q_{rad} is the radiant heat transfer from the plate surface to the surrounding test chamber.

The total heat transfer rate to the plate was equal to the measured power supply to the plate heater. The heat loss from the heated plate by radiation was calculated by assuming that the emissivity of the polished aluminum surface was 0.05, while the heat losses through the bottom and sides of the plate were determined from measurements of the heat loss from the plate when it was covered with a thick layer of Styrofoam insulation material. Using the value of the convective heat transfer rate from the plate so determined the mean Nusselt number could be determined using:

$$Nu = \frac{Q_{conv} h}{kA (T_{W_{avg}} - T_F)} \qquad (5.3)$$

A being the plate surface area. The fluid properties in the Nusselt and Rayleigh numbers are evaluated at the mean film temperature $(T_{W_{avg}} + T_F)/2$.

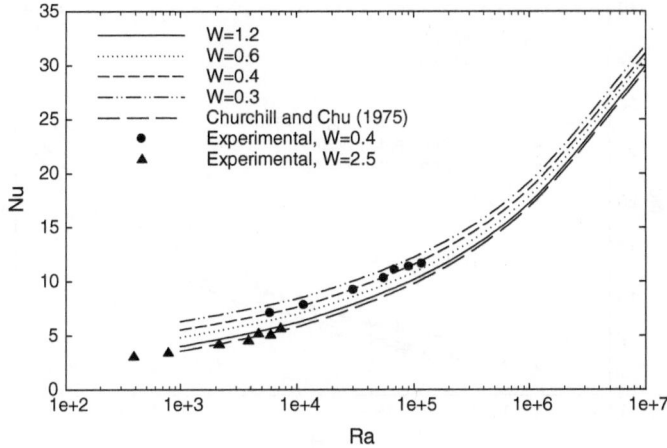

Fig. 5.4 Variation of mean Nusselt number with Rayleigh Number for various values of W for $\varphi = 0°$

During testing, a given power was supplied and the system was allowed to reach thermal equilibrium and the average surface temperature of the plate was determined from the thermocouple readings. After the data were collected, the power input to the test unit was increased to the next desired value. Again the data was collected after a steady state was obtained. The tests were carried out for a range of the heated plate's average temperatures between approximately 30 and 95 °C. The other two parameters that varied during the testing were the inclination angle, φ, and the aspect ratio of the plate, W.

5.4 Results

As mentioned before the Nusselt number value is dependent on:

1. The Rayleigh number, Ra, based on the plate height, h, and the overall temperature difference between the plate temperature and the fluid temperature.
2. The dimensionless plate width, $W = w/h$
3. The Prandtl number, Pr
4. The angle of inclination of the plate from the vertical, φ

The experimental results obtained here were essentially for $Pr = 0.7$.

Numerical and experimental variations of the mean Nusselt number with Rayleigh number for various values of the dimensionless plate width, W, for inclination angles of 0°, +45°, and −45° are shown in Figs. 5.4, 5.5, and 5.6. Also, shown in these figures are the results given by the Churchill and Chu [4] correlation equation for natural convection from a vertical wide flat plate modified to

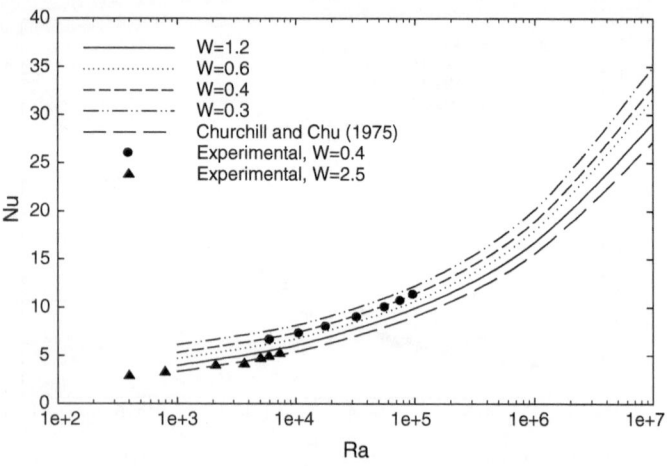

Fig. 5.5 Variation of mean Nusselt number with Rayleigh number for various values of W for $\varphi = +45°$

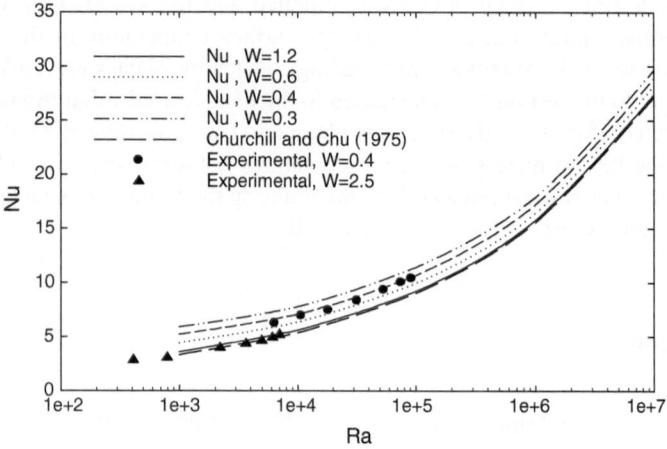

Fig. 5.6 Variation of mean Nusselt number with Rayleigh number for various values of W for $\varphi = -45°$

apply to an inclined plate by replacing the Rayleigh number, Ra, in this equation by $Ra \cos \varphi$. This correlation then gives:

$$Nu_0 = 0.68 + \left\{ \frac{0.67}{\left[1 + (0.492/\operatorname{Pr})^{9/16}\right]^{4/9}} \right\} (Ra \cos \varphi)^{0.25} \qquad (5.4)$$

which becomes for $Pr = 0.7$:

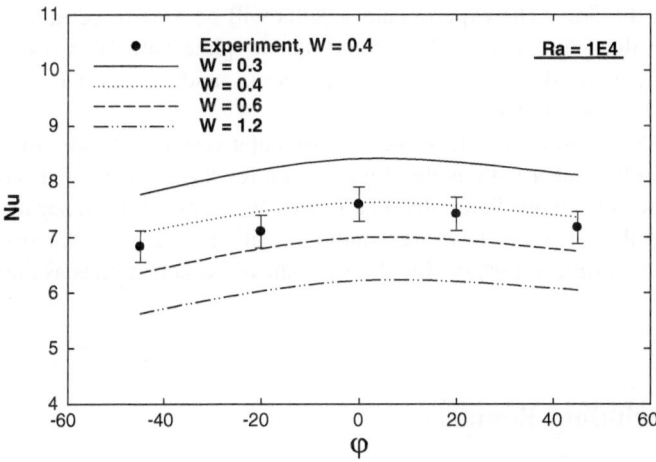

Fig. 5.7 Numerical variation of *Nu* with angle of inclination, φ, for *Ra* = 1E4 and for various values of *W*

$$Nu_0 = 0.68 + 0.513(Ra \cos \varphi)^{0.25} \qquad (5.5)$$

It will be seen from Fig. 5.4 that for $\varphi = 0°$, i.e., for a vertical plate, that, particularly at the lower Rayleigh numbers, the mean Nusselt number tends to increase with decreasing *W*. As mentioned before, the increase in the Nusselt number with decreasing *W* arises from the fact that there is an induced inflow toward the plate from the sides and this causes the heat transfer rate to be higher near the vertical sides of the plate than in the center region of the plate. At the higher values of *Ra* the edge effect become less important and as the width of the plate, *W*, increases, the results are in a good agreement with the results given by the correlation equation for wide plate.

Figures 5.5 and 5.6 give results for $\varphi = +45°$ and $\varphi = -45°$, respectively. From Fig. 5.5, it will be seen that for the upward facing heated plate at all values of *Ra* considered, the Nusselt number is higher than that given by the correlation equation, the difference increasing with decreasing *W*. The difference between the effect of *W* on the Nusselt number for the upward facing plate and the vertical plate basically arises because with the inclined plate there is a component of the buoyancy force normal to the plate surface which gives rise to pressure changes across the flow over the plate, the pressure at the surface for the case of the upward facing being lower than in the fluid outside of the flow. This pressure difference causes the flow near the edges of the plate to be different from the flow over a vertical plate. From Fig. 5.6, which shows results for the case of a downward facing plate, it will be seen that the variation of *Nu* with *Ra* is closer to that existing with a vertical plate but the effect of *W* on the results continues to be higher at larger values of *Ra* than that for the vertical plate. With the downward facing plate the pressure at the surface of the plate is higher than in the fluid outside of the flow. The experimental results will be seen to be in a good agreement with the numerical results for the narrow plate with $W = 0.4$ and for the wider plate with $W = 2.5$ under all conditions and for all inclination angles considered in this study.

Figure 5.7 shows numerical and experimental results for the variation of the mean Nusselt number with angle of inclination for $Ra = 10^4$ for four values of W. It will be seen from this figure that in all cases Nu is smaller for a negative angle of inclination than it is for the corresponding positive angle of inclination. Furthermore, the experimental results for $W = 0.4$ show the same trends as the numerical results.

5.5 Concluding Remarks

In all situations for which experimental results were obtained there was agreement between the experimental and numerical results within the experimental uncertainty. This adds further confirmation of the adequacy of the approach used in obtaining the numerical results presented in this book.

References

1. Kalendar AY, Oosthuizen PH (2011) Numerical and experimental studies of natural convective heat transfer from vertical and inclined narrow isothermal flat plates. Heat Mass Transfer 47(9):1181–1195
2. Moffat RJ (1983) Using uncertainty analysis in the planning of an experiment. J Fluids Eng, Trans ASME 107(2):173–178
3. Moffat RJ (1988) Describing the uncertainties in experimental results. Exp Therm Fluid Sci 1(1):3–17
4. Churchill SW, Chu HHS (1975) Correlating equations for laminar and turbulent free convection from a vertical plate. Int J Heat Mass Transfer 18(11):1323–1329

Chapter 6
Turbulent Natural Convective Heat Transfer from Narrow Plates

Keywords Natural convection · Turbulent flow · Transitional flow · Vertical plate · Inclined plate · Numerical

6.1 Introduction

As discussed in earlier chapters, two-dimensional natural convective heat transfer from vertical and inclined plates has been extensively studied. However, as shown in earlier chapters, in the laminar flow region when the width of the plate is relatively small compared to its height, the heat transfer rate can be considerably greater than that predicted by these two-dimensional flow results. The increase in the heat transfer rate from narrow plates relative to that from wide plates under the same conditions results from the fact that fluid flow is induced inwards near the edges of the plate and the flow near the edge of the plate is thus three dimensional. When the narrow plate is inclined, i.e., is set at an angle to the vertical, pressure changes normal to the plate surface arise and the presence of these pressure changes can alter the nature and the magnitude of the edge effects on the heat transfer rate from the plate. In many practical situations, the natural convective flow over what is effectively a narrow plate is turbulent and, hence, there exists a need to predict edge effects and heat transfer rates from narrow vertical and inclined plates in the transition region between the laminar and the turbulent flow regions and in the turbulent flow region. This is discussed in this chapter.

The heat transfer from a narrow vertical or inclined isothermal plate embedded in a plane adiabatic surface, with the adiabatic surface being in the same plane as the heated plate, has again been considered, this situation being shown previously in Chap. 1 for example. Attention has again been restricted to results for a Prandtl

P. H. Oosthuizen and A. Y. Kalendar, *Natural Convective Heat Transfer from Narrow Plates*, SpringerBriefs in Thermal Engineering and Applied Science, DOI: 10.1007/978-1-4614-5158-7_6, © The Author(s) 2013

number of 0.7, this being approximately the value existing in the application, which involved natural convection to air, that originally motivated the study on which this chapter is based.

There have been a number of studies of turbulent natural convective flow over wide flat plates, e.g., see Kitamura et al. [1], Plumb and Kennedy [2], and Schmidt and Patankar [3]. The work discussed in this chapter is based on the study described by Kalendar and Oosthuizen [4].

6.2 Solution Procedure

The flow has been assumed to be steady and fluid properties have been assumed constant, except for the density change with temperature which gives rise to the buoyancy forces, this having been treated by using the Boussinesq approach. It has also been assumed that the flow is symmetrical about the vertical center-plane of the plate. The solution has been obtained by numerically solving the full three-dimensional governing equations subject to the boundary conditions using the commercial finite volume method-based code solver, FLUENT©. The k-ε turbulent model with the full effect of buoyancy forces accounted for and with enhanced wall treatment functions has been used in obtaining the solutions. This turbulence model has, in past studies involving natural convection, been found to give moderately good predictions of when transition to turbulence occurs and of the heat transfer rate in the turbulent and transition region.

Because the flow has been assumed to be symmetrical about the vertical center-line of the plate,the solution domain used in obtaining the solution is as shown previously in Chap. 2 (Fig. 2.2). Extensive grid and convergence criterion independence testing was undertaken. This indicated that the heat transfer results presented here are to be within 1 %, independent of the number of grid points and of the convergence-criterion used. The effect of the positioning of the outer surfaces of the solution domain from the heated surface was also examined and the positions used in obtaining the results discussed here were chosen to ensure that the heat transfer results were independent of this positioning to within 1 %.

The heat transfer rate from the heated plate has been expressed in terms of the following mean and local Nusselt numbers defined as follows:

$$Nu = \frac{\bar{q}'h}{k(T_H - T_F)}, \quad Nu_y = \frac{q'y}{k(T_H - T_F)} \tag{6.1}$$

where $\bar{q}'$ and q' are the mean and local heat transfer rates per unit area. T_H and T_F are the temperatures of the heated plate surface and of the undisturbed fluid far from the plate, respectively.

An indication of the adequacy of the numerical model is given by comparing the results obtained here for the limiting case of natural convective heat transfer

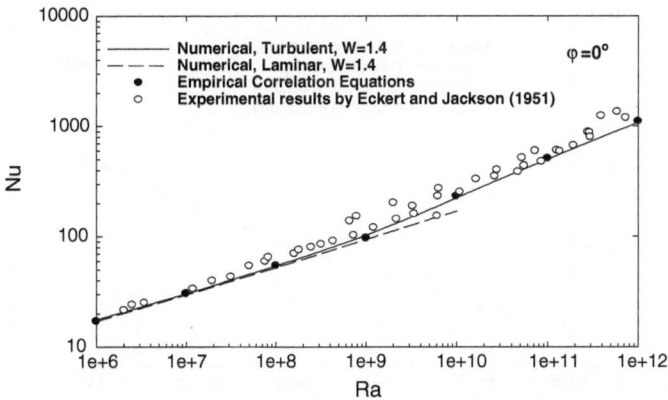

Fig. 6.1 Variation of mean Nusselt number with Rayleigh number given by numerical and experimental results and by correlation equations for natural convection from vertical flat plate when $\varphi = 0°$

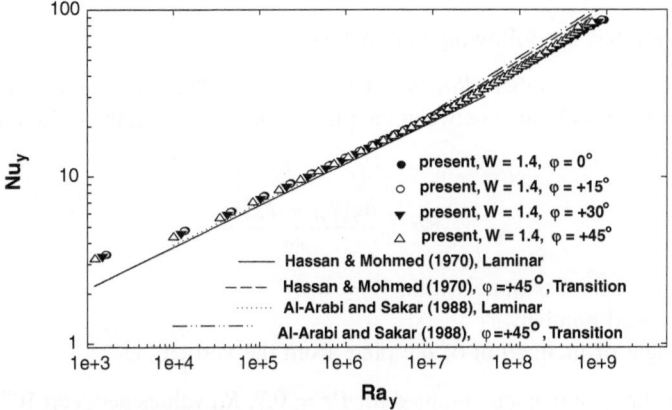

Fig. 6.2 Variation of local Nusselt number with Rayleigh number for a wide plate given by numerical results and by correlation equations for flat plates with inclination angles, φ, between $0°$ and $+45°$

from a wide plate with the results from existing experimental studies and from empirical-derived equations that are available for this flow situation in the laminar, transition, and turbulent flow regions. Such a comparison is made in Figs. 6.1 and 6.2. A consideration of these results indicates a good agreement exists between the present numerical results and the existing experimental results and with the results given by widely used correlation equations.

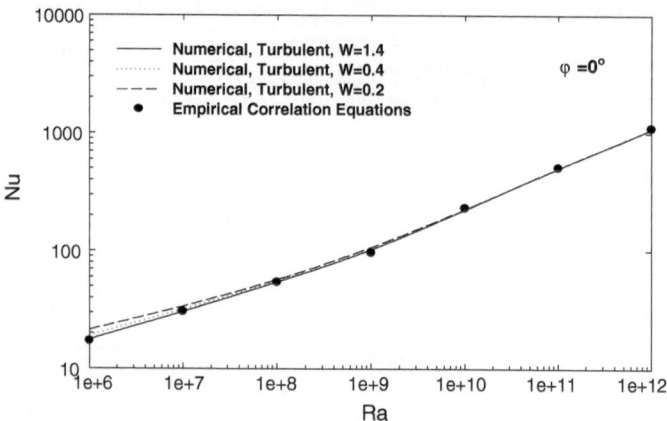

Fig. 6.3 Variation of mean Nusselt number with Rayleigh number for numerical and empirical results for natural convection from vertical flat plat when $\varphi = 0°$ and different values of W

6.3 Results

The solution has the following parameters:

1. The Rayleigh number, Ra, based on the plate "height", h, and the overall temperature difference between the plate temperature and the fluid temperature, i.e.:

$$Ra = \frac{\beta g(T_H - T_F)h^3}{\nu \alpha} \qquad (6.2)$$

2. The dimensionless plate width, $W = w/h$
3. The Prandtl number, Pr
4. The angle of inclination of the plate from the vertical, φ

Results have only been obtained for $Pr = 0.7$. Ra values between 10^6 and 10^{12}, W values between 0.2 and 1.4, and inclination angles, φ, between $-45°$ and $+45°$ have been considered.

Numerically derived variations of the mean Nusselt number with Rayleigh number for a vertical flat plate when $\varphi = 0°$ and with different values of W are shown in Fig. 6.3. This figure shows that the dimensionless width of the plate has no effect on the mean Nusselt number in the transition and turbulent flow regions; this is because in these regions the boundary layer is very thin which causes the edge effects to be negligible whereas in the laminar region the edge effects become important because the boundary layer is thicker than that in the transition and turbulent flow regions which allow an induced flow near the vertical edges and cause the mean Nusselt number to increase as the dimensionless plate width decreases.

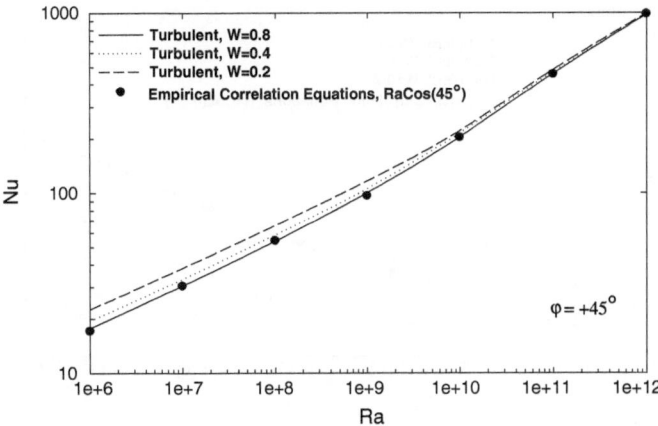

Fig. 6.4 Variation of mean Nusselt number with Rayleigh number for numerical and empirical results for natural convection from inclined flat plat when $\varphi = +45°$ and different values of W

Results given in Figs. 6.4 and 6.5 show variations of the mean Nusselt number with Rayleigh number for an inclined flat plate for $\varphi = +45°$, where the hot surface is facing up and for $\varphi = -45°$, where the hot surface is facing down, respectively, for different values of W. These figures also show the results given by the empirical vertical wide plate equation modified to apply to an inclined plate by replacing the Rayleigh number, Ra, by $Ra \cos \varphi$. Figure 6.4 shows that in the turbulent region the mean Nusselt numbers given by the modified empirical correlation equations are in a good agreement with the numerical results for all values of dimensionless plate width considered while in the laminar and transition regions the modified empirical equation results are in a good agreement with the numerical results for the larger dimensionless plate widths considered. When the heated plate surface is facing up there is a component of the buoyancy force normal to the plate surface which gives rise to pressure changes across the flow over the plate, the pressure at the surface for the case of the upward facing heated plate being lower than in the fluid outside of the flow. This pressure difference causes the flow to be induced near the edges of the plate toward the surface of the hot plate and the edge effects become more significant as the dimensionless plate width decreases and higher edge effects exist in the laminar and transition regions. For an inclined heated surface facing up, edge effects become more significant and mean Nusselt number increases as the dimensionless plate width decreases. For an inclined heated surface facing down Fig. 6.5 shows that the mean Nusselt number values given by the modified empirical correlation equations are in a good agreement with the numerical results for all values of dimensionless plate width in all flow regions considered and the mean Nusselt number increases as the dimensionless plate width decreases in the laminar flow region and with less edge effect in the transition and turbulent flow regions. With the downward facing

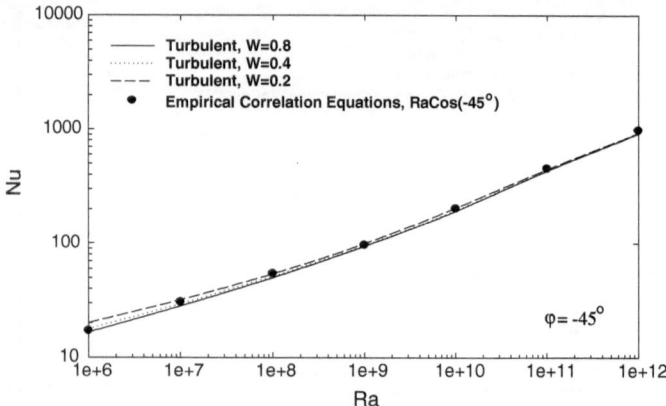

Fig. 6.5 Variation of mean Nusselt number with Rayleigh number for numerical and empirical results for natural convection from inclined flat plat when $\varphi = -45°$ and different values of W

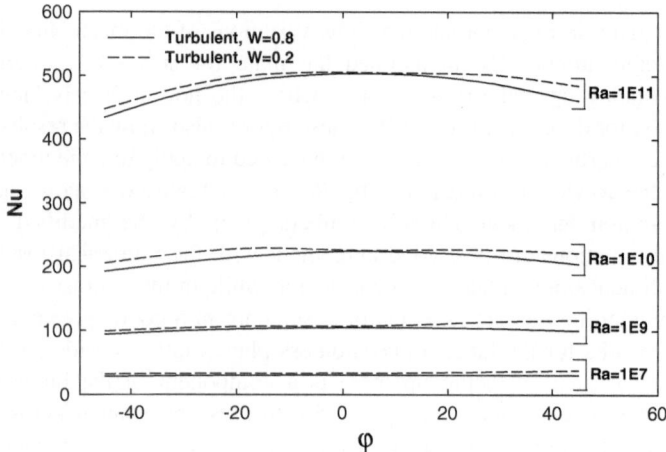

Fig. 6.6 Variation of mean Nusselt number with inclination angle for different Rayleigh number when $W = 0.2$ and 0.8

heated plate, the pressure at the surface of the plate is higher than in the fluid outside of the flow which causes an outward flow away from the heated surface.

Figure 6.6 shows the numerically predicted variation of the mean Nusselt number with angle of inclination for different values of Ra for $W = 0.8$ and for $W = 0.2$. The results given in Fig. 6.6 show that the mean Nusselt number increases as the dimensionless plate width decreases and in all cases the mean Nusselt number is smaller for a negative angle of inclination than it is for the corresponding positive angle of inclination. This difference increases as Rayleigh number increases.

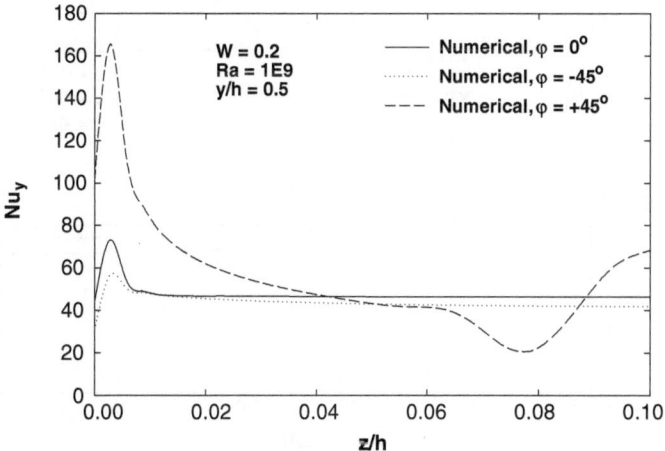

Fig. 6.7 Variation of local Nusselt number with dimensionless distance along the horizontal center line between the side edge and the vertical center line for inclination angles of $0°$, $-45°$ and $+45°$ when $W = 0.2$ and $Ra = 1E9$

To examine the differences between the edge effects for vertical and inclined flat plates facing up and facing down, the local Nusselt number along the horizontal center line of the heated plate ($y/h = 0.5$) was studied. Results given in Fig. 6.7 show that the local Nusselt number when the plate is inclined facing up is higher than that when the plate is vertical and inclined facing down near the edge of the plates , i.e., when $z/h < 0.04$. Although the buoyancy force parallel to the plate for the case of an inclined plate is lower than the buoyancy force for the vertical case, the local Nusselt number for an inclined plate facing up is higher than that for a vertical and inclined plate facing down near the edge of the plate. When the heated plate is facing up the edge effect covers more area than when the heated plate is facing down. This is because when the plate is inclined at an angle to the vertical where the hot surface is facing up a buoyancy component exists in the direction away from the surface which causes the pressure at the heated plate surface to be lower than that in the surrounding fluid and an inward fluid flow with a higher temperature gradient exists near the vertical edges induced toward the center of the heated plate thus giving rise to a higher heat flux than in the vertical and inclined heated plate facing downwards cases.

Figure 6.8 shows the variation of the local Nusselt number with dimensionless distance z/h for inclination angles of $0°$, $-15°$, $-30°$, and $-45°$. It will be seen that the local Nusselt number has higher values near the edge of the heated plate and that the values decrease as z/h increases. It will also be noted that the more negative the inclination angle (heated plate facing downwards) the lower the values of the local Nusselt number.

Figure 6.9 shows the variation of the local Nusselt number with dimensionless distance z/h for an inclination angle of $0°$, $+15°$, $+30°$, and $+45°$. This figure shows that, when the hot surface is facing upwards, as the angle of inclination increases

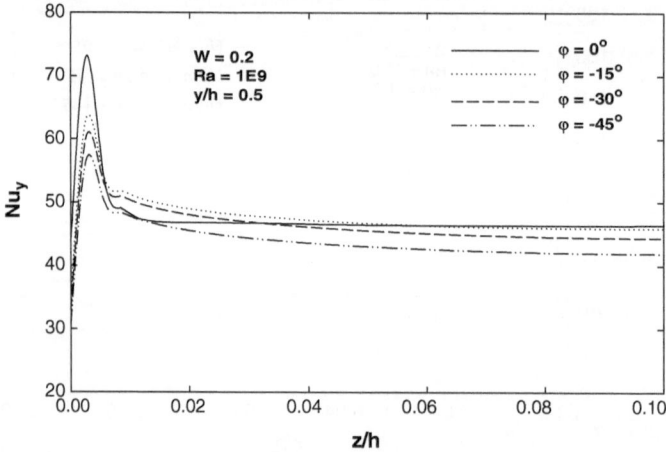

Fig. 6.8 Variation of local Nusselt number with dimensionless distance along the horizontal center line between side edge and vertical center line ($y/h = 0.5$) for inclination angle of $0°$, $-15°$, $-30°$ and $-45°$ when $Ra = 1E9$ and $W = 0.2$

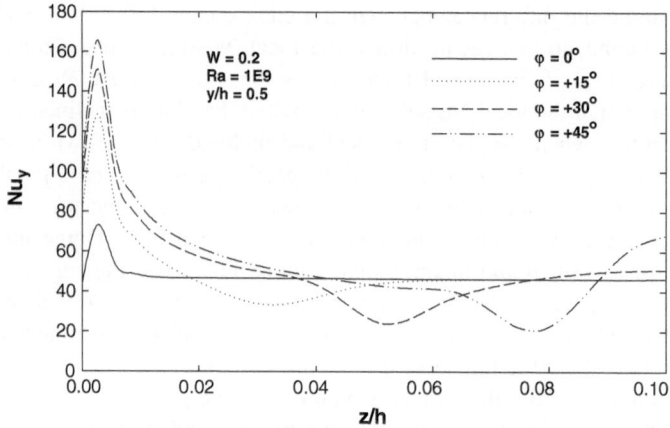

Fig. 6.9 Variation of local Nusselt number with dimensionless distance along the horizontal center line between side edge and vertical center line ($y/h = 0.5$) for inclination angle of $0°$, $+15°$, $+30°$, and $+45°$ when $Ra = 1E9$ and $W = 0.2$

the local heat transfer rate increases near the edge of the hot surface when $z/h < 0.04$. Furthermore, as the angle of inclination increases the minimum local heat transfer moves toward $z/h = 0.1$, i.e., toward the center of the heated plate because as the inclination angle increases the pressure at the heated surface increases which increases the induced flow from the edge of the plate toward the center of the heated plate. When the heated plate is facing up the buoyancy component exists in the direction away from the surface that pushes the flow away

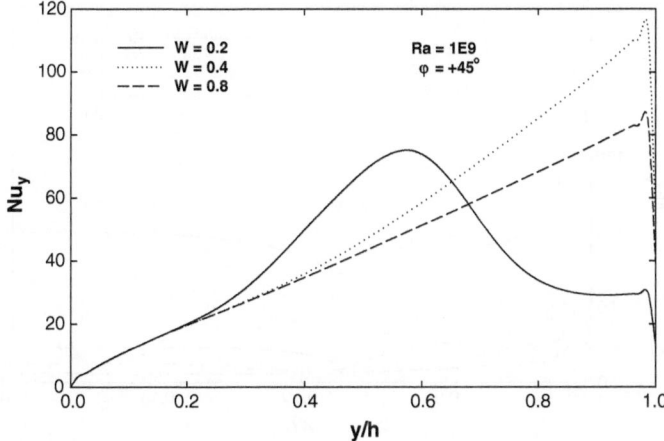

Fig. 6.10 Variation of local Nusselt number with dimensionless distance along the vertical center line between side edge and vertical center line when $W = 0.2$, 0.4, and 0.8 for an inclination angle of +45° and for $Ra = 1E9$

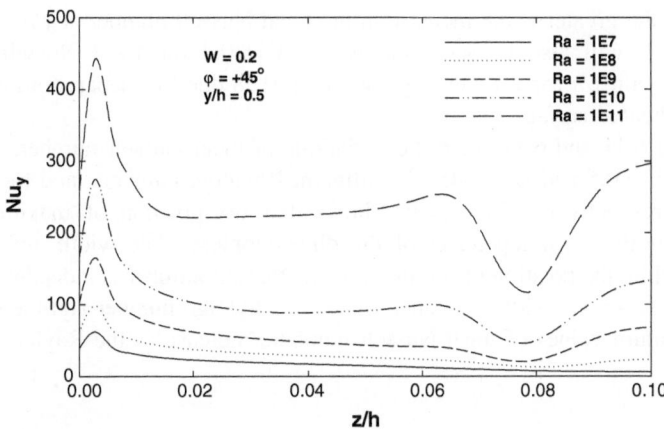

Fig. 6.11 Variation of local Nusselt number with dimensionless distance along the horizontal center line between side edge and vertical center line ($y/h = 0.5$) for inclination angle of +45° and $W = 0.2$ for different Ra

from the heated surface which causes the flow to separate from the heated plate and leading to a low heat transfer rate region.

Figure 6.10 shows the variation of the local Nusselt number with dimensionless distance y/h along the vertical center line when $W = 0.1$, 0.2, and 0.4 for an inclination angle of +45°. It will be seen that as the dimensionless plate width

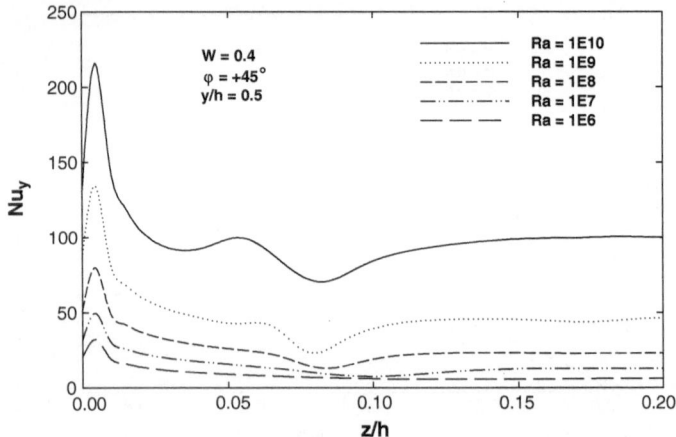

Fig. 6.12 Variation of local Nusselt number with dimensionless distance along the horizontal center line between side edge and vertical center line ($y/h = 0.5$) for inclination angle of $+45°$ and $W = 0.4$ for different Ra

decreases the greater is the increase in the local Nusselt number at y/h values less than 0.5. It will also be seen that when $W = 0.2$ the local Nusselt Number decreases with y/h for y/h values greater than 0.58 due to a separation of the flow from the heated surface.

Figures 6.11 and 6.12 show the variations of local Nusselt number, Nu_y, with z/h for $y/h = 0.5$ and $\varphi = +45°$ for different Rayleigh numbers, and for $W = 0.2$ and 0.4, respectively. The results shows that the position of maximum local Nusselt number is independent of the dimensionless plate width and Rayleigh number while the position of minimum local Nusselt number is independent of the dimensionless plate width at higher values of Rayleigh number and the minimum and maximum values of local Nusselt numbers increase as the Rayleigh number increases.

6.4 Conclusions

The results obtained in the study described in this chapter indicate that:

1. For the conditions considered the dimensionless width of the vertical flat plate has no effect on the mean Nusselt number in the transition and turbulent regions.
2. The dimensionless plate width has no effect on the mean Nusselt number for negative angle of inclination, i.e., for the case where hot surface is facing down, in the turbulent and transition flow regions

3. The dimensionless plate width has no effect on the mean Nusselt number for positive angle of inclination, i.e., where hot surface is facing up, in the turbulent flow region but the dimensionless plate width effects do start to become significant in the transition flow region.

4. The modified empirical equations for an inclined plate obtained by replacing the Rayleigh, Ra, by $Ra \cos \varphi$, in the empirical equations for a vertical plate gives a good prediction of the mean Nusselt number in the transition and turbulent regions for all dimensionless plate width for the upward facing plate case.

5. The mean Nusselt number is smaller for a negative angle of inclination than it is for the corresponding positive angle of inclination and this difference increases as the Rayleigh number increases.

References

1. Kitamura K, Chen X-A, Kimura F (2001) Turbulent transition mechanisms of natural convection over upward-facing horizontal plates. JSME Int J Ser B (Fluids Therm Eng) 44(1): 90–8

2. Plumb OA, Kennedy LA (1977) Application of a K-epsilon turbulence model to natural convection from a vertical isothermal surface. J Heat Transf 99(Ser C)(1):79–85

3. Schmidt RC, Patankar SV (1991) Simulating boundary layer transition with low-Reynolds-number k-turbulence models. 1. An evaluation of prediction characteristics. Trans ASME J Turbomach 113(1):10–17

4. Kalendar AY, Oosthuizen PH (2010) A numerical study of natural convective heat transfer from vertical and inclined narrow isothermal flat plates in the transition and turbulent flow regions. In: Proceedings of 2010 CSME Forum, Victoria, BC

Chapter 7
Natural Convective Heat Transfer from Two Adjacent Narrow Plates

Keywords Natural convection · Two adjacent plates · Narrow plates · Inclined plates · Isothermal plates · Numerical · Empirical equations

7.1 Introduction

In this chapter, numerical studies of the interaction of the natural convective flows over two adjacent vertical and inclined narrow isothermal flat plates in the laminar flow region are discussed. Two cases have been considered. In one case, the plates are horizontally adjacent to each other, the plates being horizontally separated (see Fig. 7.1) while in the other case, one plate is symmetrically placed above the other plate (see Fig. 7.2) the plates being vertically separated. Attention is given to the effects of the inclination angle of the two heated plates to the vertical, to the effects of the vertical and horizontal dimensionless gap between the heated plates and to the effects of the dimensionless plate width on the mean heat transfer rates from the two heated plates for a wide range of Rayleigh numbers.

The concern here is again with plates that are relatively narrow, i.e., that have relatively low width, w, to height, h, ratios. When the plate is narrow, the heat transfer rate from the plate can be, as discussed in earlier chapters of this book, considerably greater than that predicted by two-dimensional flow results. The increase in the heat transfer rate from narrow plates relative to that from wide plates under the same conditions results from the fact that fluid flow is induced inwards near the edges of the plate and the flow near the edge of the plate is, thus, three-dimensional. When the narrow plate is inclined, i.e., is set at an angle to the vertical, pressure changes normal to the plate surface arise and the presence of these pressure changes can alter the nature and the magnitude of the edge effects on the heat transfer rate from the plate.

P. H. Oosthuizen and A. Y. Kalendar, *Natural Convective Heat Transfer from Narrow Plates*, SpringerBriefs in Thermal Engineering and Applied Science, DOI: 10.1007/978-1-4614-5158-7_7, © The Author(s) 2013

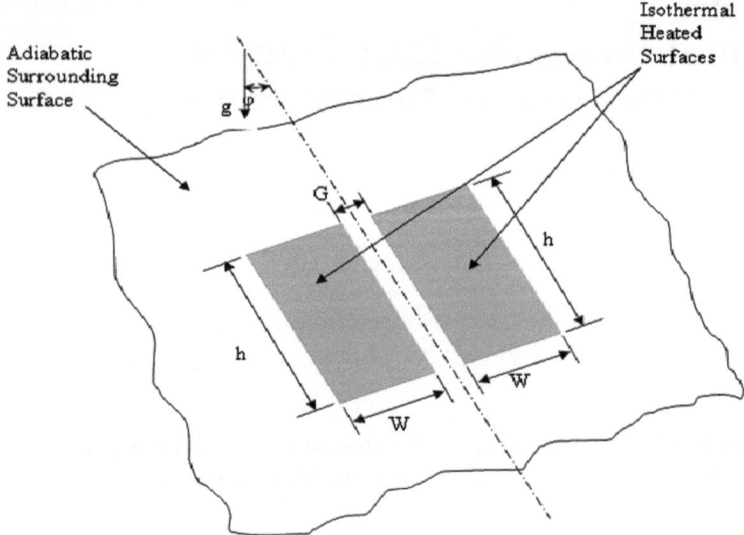

Fig. 7.1 Horizontally adjacent plate flow situation considered

Fig. 7.2 Vertically adjacent plate flow situation considered

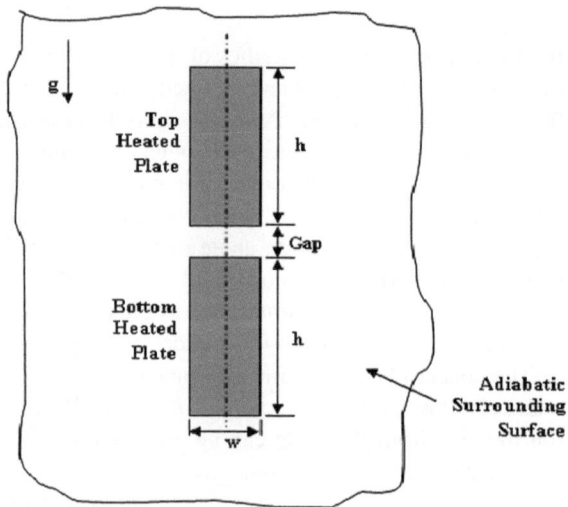

The present work arises from the fact that when there are two adjacent narrow flat plates with a relatively small gap between the plates the flow near the adjacent plates alters the nature of the flow compared to that over a single narrow plate and this can lead to a significant change in the mean heat transfer rate compared to that from a single isolated plate under the same conditions.

Situations that can be approximately modeled as two adjacent narrow heated plates that are either vertical or inclined at an angle to the vertical do occur in

Fig. 7.3 Definition of positive and negative angles of inclination

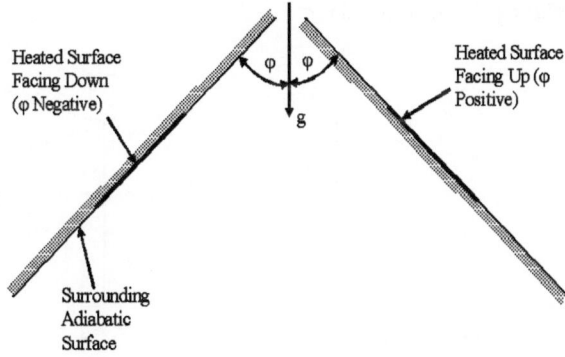

practical situations such as, for example, in some situations involving the cooling of electronic and electrical equipment. Thus, there appeared to be a need to be able to predict heat transfer rates in such adjacent narrow plate situations and this is the reason for discussing numerical results for this flow situation.

The work discussed in this chapter is largely based on that described in [1–3].

7.2 Heat Transfer from Two Horizontally Adjacent Isothermal Plates

Consideration in this section will be given to the heat transfer from two relatively narrow horizontally adjacent isothermal plates of the same size that are embedded in a plane adiabatic surface, the surface of the adiabatic surface being in the same plane as the surfaces of the heated plates. The two plates have the same uniform surface temperature and they are aligned with each other as shown in Fig. 7.1. The plates are separated from each other by a relatively small gap. The plates are, in general, inclined at an angle to the vertical (Fig. 7.3).

Results have been obtained for a relatively wide range of Rayleigh numbers and dimensionless plate widths and plate separation gaps for both positive and negative inclination angles, i.e., for both the case where the heated plates are facing up and for the case where the heated plates are facing down, these two cases being illustrated in Fig. 7.3. As previously stated, attention has been restricted to results for a Prandtl number of 0.7.

As discussed in the literature review given in Chap. 1, studies of natural convective heat transfer from arrays of heated elements with interaction of the flows from these elements are available. However, none of these studies provides much information about the effect of an adjacent element on the edge effect when narrow plates are involved.

Fig. 7.4 Solution domain

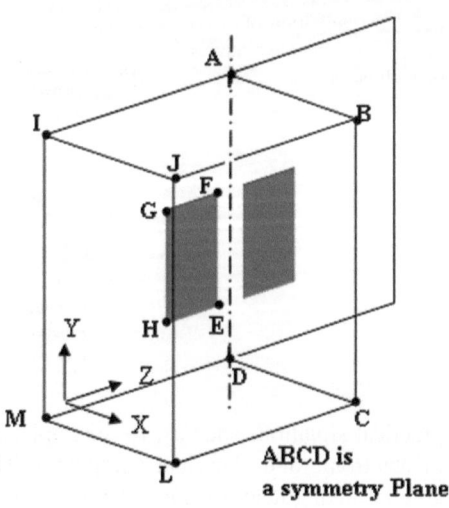

ABCD is
a symmetry Plane

7.2.1 Solution Procedure

As with most of the other numerical studies discussed in this book, the flow has been assumed to be steady and laminar. Fluid properties have been assumed constant except for the density change with temperature which gives rise to the buoyancy forces, this having been treated by using the Boussinesq approach. It has also been assumed that the flow is symmetric about the vertical center-plane between the plates (the plane ABCDA in Fig. 7.4.)

The solution has again been obtained by numerically solving the full three-dimensional form of the governing equations. These equations have been written in terms of dimensionless variables using the height, h, of the heated plate as the length scale and the overall temperature difference $(T_H - T_F)$ as the temperature scale, with T_H being the temperature of the plate surfaces and T_F being the fluid temperature far from the plate. The definitions of these dimensionless variables and the dimensionless equations that result from the use of these variables are the same as those given in earlier chapters.

The solution domain used in obtaining the solution is shown in Fig. 7.4.

Considering the surfaces shown in Fig. 7.4, the assumed boundary conditions on the solution, in terms of the dimensionless variables are, since flow symmetry is being assumed:

$$\text{FEHG}: \quad U_X = 0, U_Y = 0, U_Z = 0, \theta = 1$$

$$\text{ADMI except for FEHG}: \quad U_X = 0, U_Y = 0, U_Z = 0, \frac{\partial \theta}{\partial X} = 0$$

$$\text{BCLJ}: \quad U_Y = 0, U_Z = 0, \theta = 0$$

$$\text{JLMI}: \quad U_X = 0, U_Y = 0, \theta = 0$$

$$\text{DCLM}: \quad U_X = 0, U_Z = 0, \theta = 0$$

$$\text{ABCD}: \quad U_Z = 0, \frac{\partial U_Y}{\partial Z} = 0, \frac{\partial U_X}{\partial Z} = 0, \frac{\partial \theta}{\partial Z} = 0$$

$$\text{ABJI}: \quad P = 0$$

The mean heat transfer rate from the heated plate has again been expressed in terms of the following mean Nusselt number:

$$Nu = \frac{\bar{q}' h}{k(T_H - T_F)} \tag{7.1}$$

where $\bar{q}'$ is the mean heat transfer rate per unit area from the heated plate.

Subject to the boundary conditions discussed above, the dimensionless governing equations have been numerically solved using the commercial finite-volume solver FLUENT. Extensive grid and convergence criterion independence testing was undertaken. This indicated that the heat transfer results presented here are to within 1 % independent of the number of grid points and of the convergence criterion used. The effect of the positioning of the outer surfaces of the solution domain (i.e., surfaces BCLJ, JLMI, DCLM, and ABJI in Fig. 7.4) from the heated surface was also examined and the positions used in obtaining the results discussed here were chosen to ensure that the heat transfer results were independent of this positioning to within 1 %.

7.2.2 Results

The solution has the following parameters:

1. The Rayleigh number, Ra, based on the plate height, h, and the overall temperature difference between the plate temperature and the undisturbed fluid temperature.
2. The dimensionless plate width, $W = w/h$
3. The Prandtl number, Pr
4. The angle of inclination of the plate from the vertical, φ
5. The dimensionless gap between the two adjacent plates, $HGap = G/h$

Results were obtained for $Pr = 0.7$, Ra values between 10^3 and 10^7, W values between 0.15 and 0.6, inclination angles, φ, of between $-45°$ and $+45°$, and dimensionless gaps between the adjacent plates, $HGap$, of from 0 to 0.2.

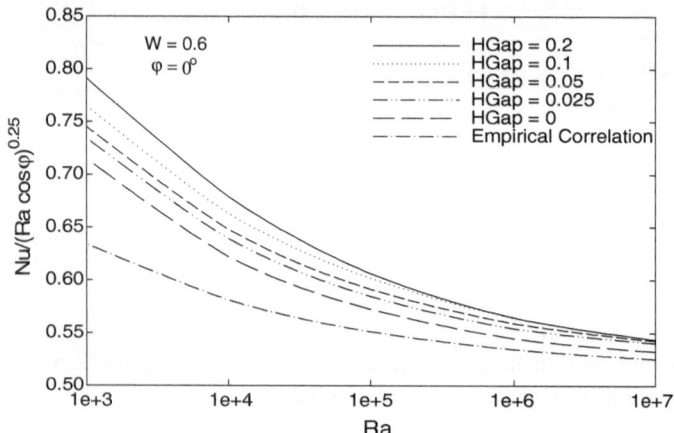

Fig. 7.5 Variation of $Nu/(Ra \cos \varphi)^{0.25}$ with Rayleigh number for various values of $HGap$ for $W = 0.6$ and $\varphi = 0°$

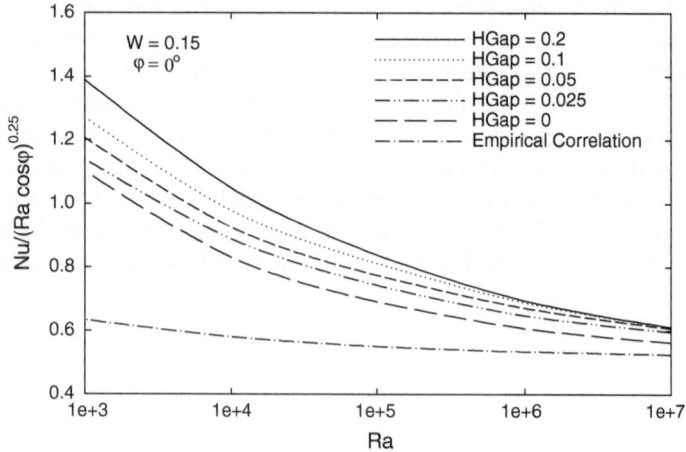

Fig. 7.6 Variation of $Nu/(Ra \cos \varphi)^{0.25}$ with Rayleigh number for various values of $HGap$ for $W = 0.15$ and $\varphi = 0°$

The effect of the dimensionless plate width on the mean Nusselt number is illustrated by the variations of the mean Nusselt number with Rayleigh numbers for various values of the dimensionless gap between the plates for dimensionless plate widths, W, of 0.6 and 0.15 for an inclination angle of 0° shown in Figs. 7.5 and 7.6. The empirical correlation shown in these figures is the based on the standard Churchill and Chu correlation equation [4] for natural convection from a wide vertical flat plate that was discussed in earlier chapters. It has again been modified to apply to an inclined plate by replacing the Rayleigh number, Ra, by $Ra \cos\varphi$. This correlation equation then gives:

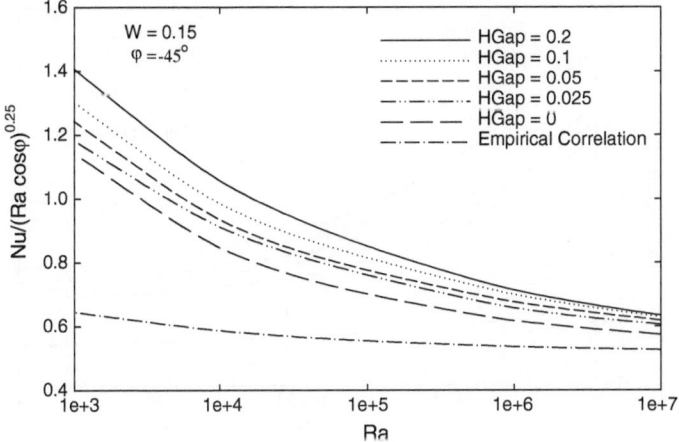

Fig. 7.7 Variation of $Nu/(Ra \cos \varphi)^{0.25}$ with Rayleigh number for various values of $HGap$ for $W = 0.15$ and $\varphi = +45°$

$$Nu_0 = 0.68 + \left\{ \frac{0.67}{\left[1 + (0.492/\,\mathrm{Pr})^{9/16} \right]^{4/9}} \right\} (Ra \cos \varphi)^{0.25} \qquad (7.2)$$

which becomes for $Pr = 0.7$:

$$Nu_0 = 0.68 + 0.513(Ra \cos \varphi)^{0.25} \qquad (7.3)$$

The results given in Figs. 7.5 and 7.6 show how the Nusselt number increases with decreasing dimensionless plate width and how it decreases with decreasing dimensionless gap size. The results also show that $HGap$ has a higher effect at lower values of Rayleigh numbers. As the dimensionless plate width increases and the dimensionless gap decreases to zero, the two adjacent plates become a single wide plate and the Nusselt number becomes closer to the empirical correlation for a wide plate particularly at higher Rayleigh numbers.

The effects of the angle of inclination on the mean Nusselt number are illustrated by the results shown in Figs. 7.6, 7.7 and 7.8 which show the variations of $Nu/(Ra \cos \varphi)^{0.25}$ with Ra for various values of the dimensionless gap, $HGap$, for inclination angles of 0°, +45°, and −45° for a dimensionless plate width of 0.15. These results show how, as a result of the edge effects, in all cases $Nu/(Ra \cos\varphi)^{0.25}$ increases with increasing $HGap$ and also increases with decreasing Ra and W. In addition, the results show that the gap between two adjacent hot plates causes different interaction effects for an inclined plate than for a vertical plate and that the interaction effect is different when the plate is facing down than when it is facing up.

The effects discussed above are further illustrated by the results shown in Figs. 7.9, 7.10, and 7.11. It will be seen from Fig. 7.9 that for $\varphi = 0°$, i.e., for

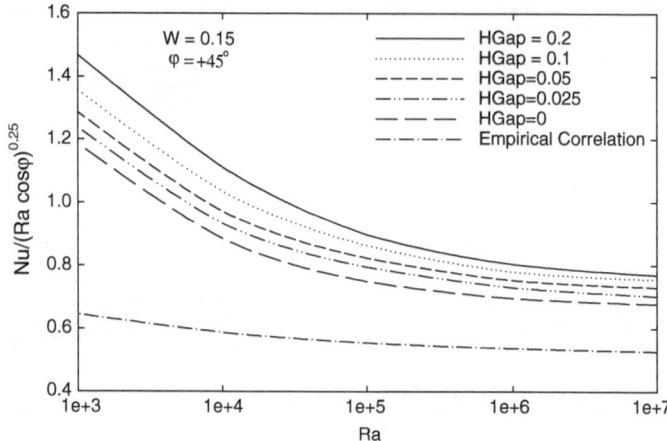

Fig. 7.8 Variation of $Nu/(Ra \cos \varphi)^{0.25}$ with Rayleigh number for Various values of *HGap* for $W = 0.15$ and $\varphi = +45°$

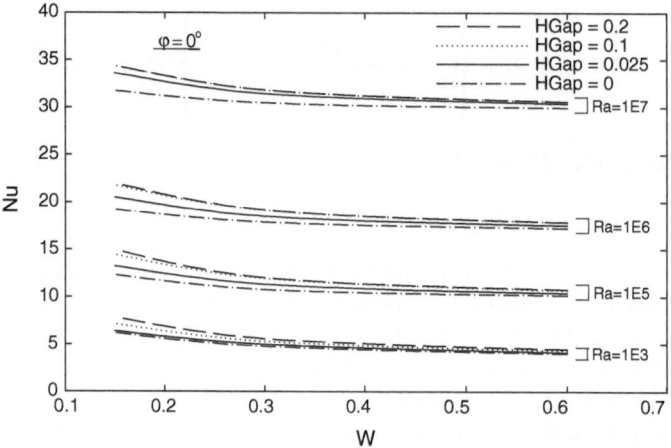

Fig. 7.9 Variation of mean Nusselt number with dimensionless plate width *W* for various values of *HGap*, and *Ra* for $\varphi = 0°$

a vertical plate, the mean Nusselt number tends to increase with decreasing *W* and increasing dimensionless gap *HGap* between two plates. As mentioned before, the increase in the Nusselt number with decreasing *W* arises from the fact that there is an induced inflow toward the plate from the sides and this causes the heat transfer rate to be higher near the vertical sides of the plate than it is in the center region of the plate. When the two heated plates are closer to each other, the interaction between the flows over the two plates increases and the temperature gradient between the two plates decreases which reduces the magnitude of the edge effect

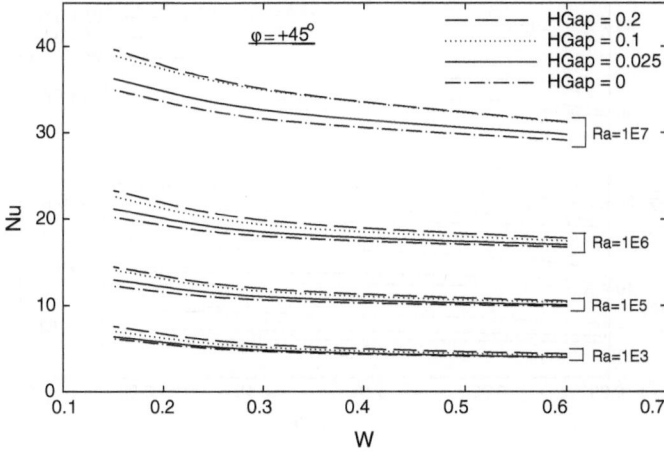

Fig. 7.10 Variation of mean Nusselt number with dimensionless plate width W for various values of *HGap* and *Ra* for $\varphi = +45°$

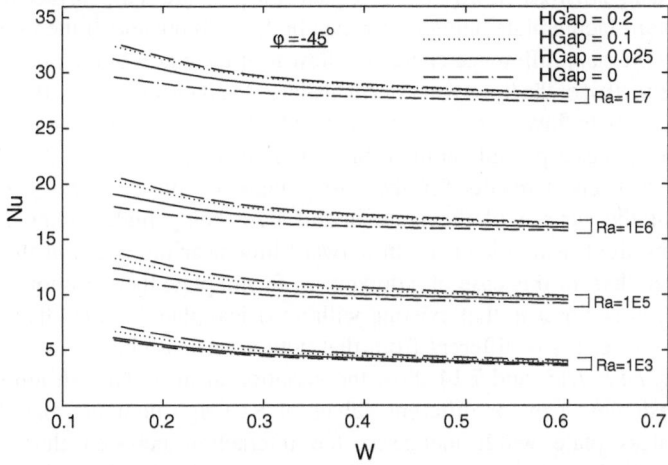

Fig. 7.11 Variation of mean Nusselt number with dimensionless plate width W for various values of *HGap*, and *Ra* for $\varphi = -45°$

from one vertical side of each plate. This effect is especially significant at low Rayleigh numbers.

From Fig. 7.10, it will be seen that for the upward facing heated plates at all values of W considered the mean Nusselt number increases with increasing *HGap* especially at the higher Rayleigh numbers, the difference increasing with decreasing W. The variation between the effect of W on the Nusselt number for the upward facing plate and for the vertical plate basically arises because with the inclined plate there is a component of the buoyancy force normal to the plate surface which gives rise to

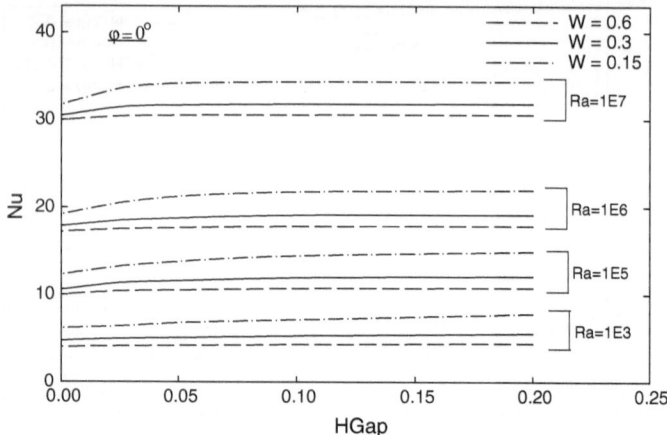

Fig. 7.12 Variation of Mean Nusselt number with dimensionless gap *HGap* for various values of *W* and Rayleigh number for $\varphi = 0°$

pressure changes across the flow over the plate, the pressure at the surface for the case of the upward facing plate being lower than in the fluid outside of the flow over the plate. This pressure difference causes the flow near the edges of the plate to be very different from that of the flow over a vertical plate. As *HGap* decreases, the interaction between two plate flows become significant which affects the flow near the vertical edges of the plates especially at high Rayleigh numbers.

Figure 7.11 gives results for the case where the plate is facing down, the pressure at the surface of the plate in this case being higher than that in the surrounding fluid, which leads to an outward flow near the edges of the plates. It will be seen that for this case of a downward facing plate the variation of *Nu* with *W* and *HGap* is closer to that existing with a vertical plate but the effect of *W* and *HGap* on the results is different from that with a vertical plate.

Figures 7.12, 7.13, and 7.14 show the variation of mean Nusselt numbers with the dimensionless gap for different values of Rayleigh numbers and *W*. As the dimensionless plate width increases, the interaction between the two plates becomes insignificant with decreasing dimensionless gap for all Rayleigh numbers and inclination angles considered here. This is because for the wider plate, i.e., with large *W*, the edge effect becomes negligible. As the dimensionless plate width decreases, the interaction between the two plates becomes significant with decreasing dimensionless gap for all Rayleigh numbers and inclination angles considered in this study. This is because for the narrow plate the edge effects are significant. When the plate is facing up, i.e., where the inclination angle is positive, at high Rayleigh numbers the interaction between the two plates becomes significant as the dimensionless plate width decreases.

Figure 7.15 shows the variation of the Nusselt number with angle of inclination for three values of *Ra* and for three values of *HGap* for *W* = 0.3. It will be seen from this figure that in all cases *Nu* is smaller for a negative angle of inclination

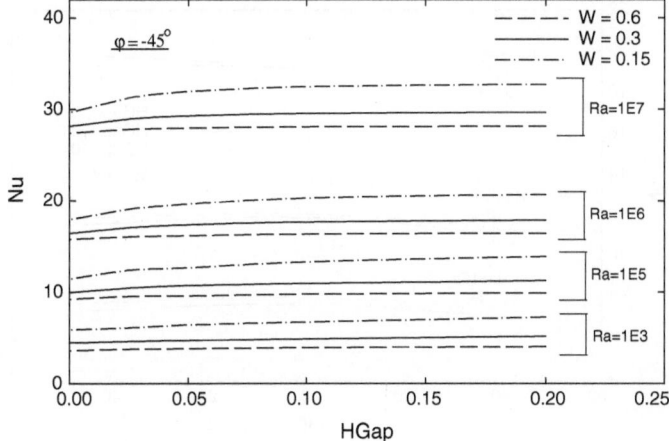

Fig. 7.13 Variation of mean Nusselt number with dimensionless gap *HGap* for various values of *W* and Rayleigh number for $\varphi = -45°$

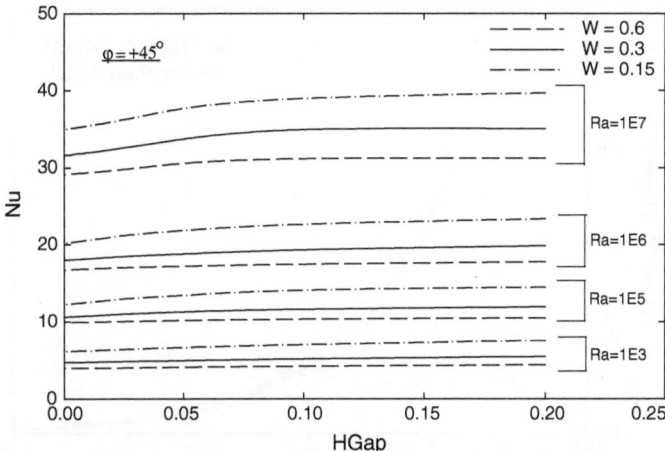

Fig. 7.14 Variation of mean Nusselt number with dimensionless gap *HGap* for various values of *W* and Rayleigh number ford $\varphi = +45°$

than it is for the corresponding positive angle of inclination. Also the *Nu* value decreases with decreasing *HGap* particularly at low Rayleigh numbers.

Because of the assumed form of the variation of *Nu* when edge effects are negligible, it has been assumed in attempting to correlate the results for the vertical plate case that the Nusselt number for a plate of dimensionless width *W* is given by an equation of the form:

$$\frac{Nu}{Ra^{0.25}} = \frac{Nu_0}{Ra^{0.25}} + \text{function } (Ra, W, HGap) \tag{7.4}$$

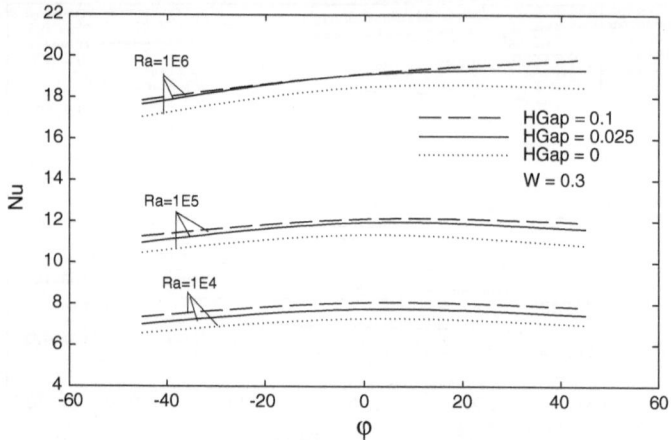

Fig. 7.15 Variation of *Nu* with angle of inclination, *φ* for various values of *HGap* and *Ra* for $W = 0.3$

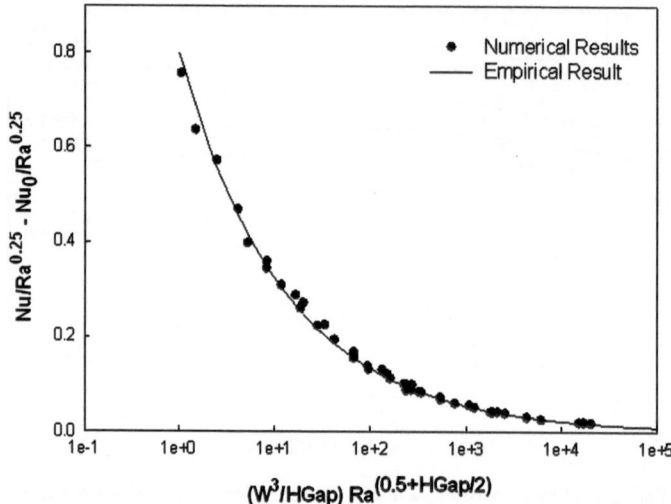

Fig. 7.16 Comparison of the results given by correlation equation for two adjacent narrow vertical flat plates with the numerical results

Using an equation of this form, it has been found that the present numerical results for a vertical plate when $HGap > 0$ can be approximately described by:

$$\frac{Nu}{Ra^{0.25}} - \frac{Nu_0}{Ra^{0.25}} = \frac{0.8}{\left[\left(\dfrac{W^3}{HGap}\right)Ra^{\left(0.5+\frac{HGap}{2}\right)}\right]^{0.39}} \tag{7.5}$$

Fig. 7.17 Solution domain

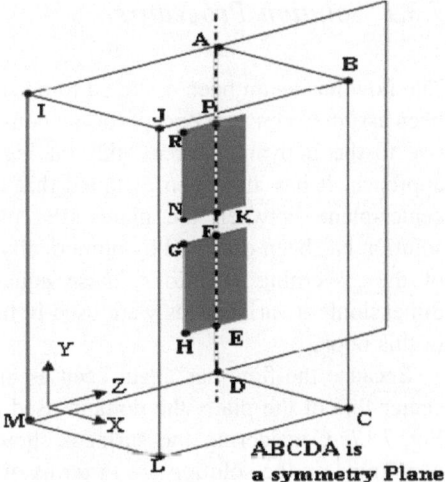

ABCDA is
a symmetry Plane

A comparison of the results given by equation (7.5) with the numerical results is shown Fig. 7.16. It will be seen that equation (7.5) describes the computed numerical results to an accuracy of better than 5 %.

7.3 Heat Transfer from Two Vertically Spaced Isothermal Plates

The heat transfer from two vertically separated narrow plates of the same size that are embedded in a plane adiabatic surface, such that the adiabatic surface is in the same plane as the surfaces of the heated plate will next be considered. The two plates have the same surface temperature and the vertical edges of the top and bottom plates are aligned. The plates are separated from each other by a vertical gap. The situation considered is, therefore, as shown in Fig. 7.17. Results have been obtained for a relatively wide range of Rayleigh numbers, dimensionless plate gaps, and dimensionless plate widths. Attention has again been restricted to results for a Prandtl number of 0.7, this being approximately the value existing in applications that involve the flow of air.

The present work arises from the fact that when there are two vertically separated flat plates with a gap between the plates, the interaction of the flow from the lower plate with that over the upper plate can alter the nature and the magnitude of the edge effects. Since situations that can be approximately modeled as two vertically separated narrow heated plates do occur in practical situations, for example in some situations involving the cooling of electronic and electrical equipment, there does therefore appear to be a need to be able to predict heat transfer rates from vertically separated narrow plates. It is for this reason that this section is included in the book.

7.3.1 Solution Procedure

The flow has again been assumed to be steady and laminar. Fluid properties have been assumed constant except for the density change with temperature which gives rise to the buoyancy forces, this having been treated by using the Boussinesq approach. It has also been assumed that the flow is symmetric about the vertical center-plane between the plates (the plane ABCDEFKPA in Fig. 7.17). The solution has been obtained by numerically solving the full three-dimensional form of the governing equations, these equations being written in terms of same dimensionless variables as were used in the work described in the earlier chapters in this book.

Because the flow has again been assumed to be symmetric about the vertical center-line of the plate, the domain used in obtaining the solution is as shown in Fig. 7.17. Considering the surfaces shown in Fig. 7.17, the assumed boundary conditions on the solution are in terms of the dimensionless variables, since flow symmetry is being assumed:

$$\text{FEHG and PKNR}: \quad U_X = 0, U_Y = 0, U_Z = 0, \theta = 1$$

$$\text{ADMI except for FEHG and PKNR}: \quad U_X = 0, U_Y = 0, U_Z = 0, \frac{\partial \theta}{\partial X} = 0$$

$$\text{BCLJ}: \quad U_Y = 0, U_Z = 0, \theta = 0$$

$$\text{JLMI}: \quad U_X = 0, U_Y = 0, \theta = 0$$

$$\text{DCLM}: \quad U_X = 0, U_Z = 0, \theta = 0$$

$$\text{ABCD}: \quad U_Z = 0, \frac{\partial U_Y}{\partial Z} = 0, \frac{\partial U_X}{\partial Z} = 0, \frac{\partial \theta}{\partial Z} = 0$$

$$\text{ABJI}: \quad P = 0$$

The mean heat transfer rate from the heated plate has again been expressed in terms of the following mean Nusselt number:

$$Nu = \frac{\bar{q}' h}{k(T_H - T_F)} \tag{7.6}$$

where $\bar{q}'$ is the mean heat transfer rate per unit area from the heated plate.

The dimensionless governing equations subject to the boundary conditions discussed above have been numerically solved using the commercial finite-volume solver FLUENT©. Extensive grid and convergence criterion independence testing was undertaken. This indicated that the heat transfer results presented here are to within 1 % independent of the number of grid points and of the convergence criterion used. The effect of the positioning of the outer surfaces of the solution domain (i.e., surfaces BCLJ, JLMI, DCLM, and ABJI in Fig. 7.17) from the heated surface was also examined and the positions used in obtaining the results discussed here were chosen to ensure that the heat transfer results were independent to within 1 % of this positioning.

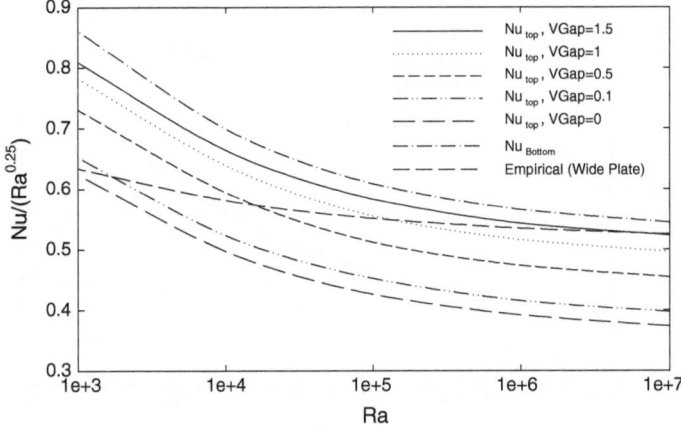

Fig. 7.18 Variation of $Nu/Ra^{0.25}$ for the top and bottom plates with Rayleigh number for various values of dimensionless gap, *VGap*, for $W = 0.6$

7.3.2 Results

The solution has the following parameters:

1. The Rayleigh number, *Ra*, based on the plate "height", *h*, and the overall temperature difference between the plate temperature and the undisturbed fluid temperature.
2. The dimensionless plate width, $W = w/h$
3. The Prandtl number, *Pr*.
4. The dimensionless gap between the two vertically separated plates, $VGap = Gap/h$

Results for $Pr = 0.7$, *Ra* values between 10^3 and 10^7, *W* values between 0.2 and 0.6 and dimensionless gaps between the two vertically separated plates, *VGap,* from 0 to 1.5 will be discussed here.

The effect of the dimensionless vertical gap between two plates on the mean Nusselt numbers for the two plates is illustrated by the results given in Figs. 7.18, 7.19, and 7.20. The empirical correlation shown in these figures is again that given by Churchill and Chu [4] for natural convection from a wide vertical flat plate. This correlation equation was discussed in the preceding section.

The effects of the dimensionless gap between two heated plates and of the dimensionless plate width, *W*, on the mean Nusselt number of the top plate, Nu_{top}, and the bottom plate, Nu_{bottom}, are illustrated by the results shown in Fig. 7.18 which shows the variations of $Nu/(Ra)^{0.25}$ with *Ra* for various values of the dimensionless gap *VGap* for a dimensionless plate width $W = 0.6$. These results show how, in all cases, as a result of the interaction between the flows over the two plates, $Nu/(Ra)^{0.25}$ for the top plate increases with increasing *VGap* and also increases with decreasing *Ra*. It will also be seen from Fig. 7.18 that the mean

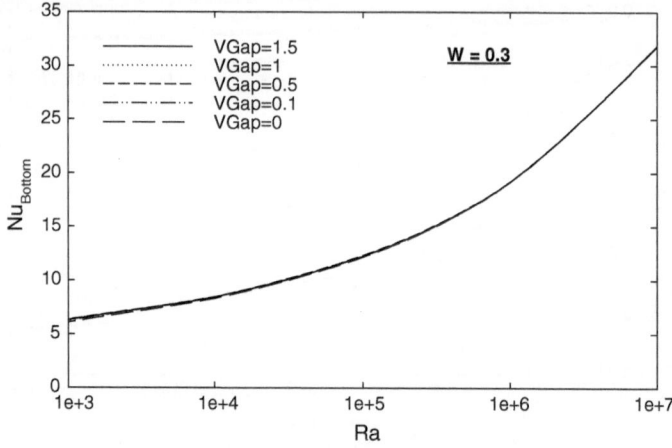

Fig. 7.19 Variation of Nu_{bottom} with Rayleigh number for various values of dimensionless gap, *VGap*, for $W = 0.3$

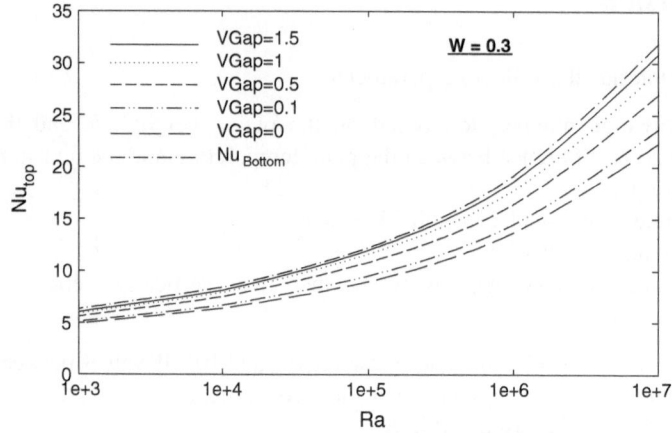

Fig. 7.20 Variation of Nu_{top} with Rayleigh number for various values of dimensionless gap, *VGap*, for $W = 0.3$

Nusselt number for the bottom plate, Nu_{bottom}, tends toward the value given by the empirical correlation equation for two-dimensional flow at high Rayleigh number but has a higher value than that given by the correlation equation at low Rayleigh number due to edge effects. Thus, as the dimensionless plate width, W, increases the mean Nusselt number for the bottom plate gets closer to the value given by the two-dimensional empirical correlation equation discussed in earlier chapters. However, the mean Nusselt number for the top plate, Nu_{top}, does not tend to the value given by the two-dimensional flow empirical equation except at the highest dimensionless gap ratio considered.

The results given in Fig. 7.19 show that at all Rayleigh numbers considered in the present study the mean Nusselt number for the bottom plate, Nu_{bottom}, is not affected by the dimensionless gap between two heated plates, $VGap$.

The results given in Fig. 7.20 show that the mean Nusselt number of the top plate, Nu_{top}, increases as the dimensionless gap between two heated plates, $VGap$, increases. It will also be seen from Fig. 7.20 that the dimensionless gap, $VGap$, has a greater effect on the mean Nusselt number at higher Rayleigh numbers than that at low Rayleigh numbers. This is because at low Rayleigh numbers edge effects are more significant than at high Rayleigh numbers and the increase in mean Nusselt number that these edge effects cause on both the upper and lower plates dominates over the effect of the interference between the flows over the two plates. It will also be seen from Fig. 7.20 that at the largest dimensionless gap between two heated plates, $VGap$, considered, i.e., 1.5, the mean Nusselt numbers for the two plates are relatively close at all Rayleigh numbers considered.

Typical variations of the mean Nusselt number for the top and bottom plates with dimensionless gap between the plates for various Rayleigh numbers for dimensionless plate widths of 0.6 and 0.3 are shown in Figs. 7.21 and 7.22. It will again be seen that the presence of the upper plate has essentially no effect on the mean Nusselt number for the lower plate. It will also be seen that at small dimensionless gap values the mean Nusselt number for the top plate is much lower than that for the bottom plate but that for dimensionless gaps greater than approximately 1.5, the mean Nusselt number for the two plates become essentially equal. It will also be seen from Figs. 7.21 and 7.22 that the mean Nusselt numbers for both the top and bottom plates increases as the dimensionless plate width, W, of the plates decreases. This is because for the narrow plate the edge effects become more significant especially at low Rayleigh number. When the two plates are closer to each other, especially at high Rayleigh numbers when the edge effect becomes less significant, the presence of the bottom plate has more effect on the top plate leading to a reduction in the heat transfer rate from the top surface as the dimensionless gap, $VGap$, decreases.

The overall mean Nusselt number based on the mean heat transfer rate averaged over both the top and bottom plates, Nu_{total}, will be considered next. It will be seen from Fig. 7.23 that this overall mean Nusselt number, Nu_{total}, increases as the dimensionless gap between two plates increases and as the dimensionless plate width, W, of the plates decreases. It will also be noted that the dimensionless gap between two plates has less effect on the overall mean Nusselt numbers, Nu_{total}, at lower Rayleigh numbers than at higher Rayleigh numbers.

A consideration of the results shown in Figs. 7.24 and 7.25 indicates that the mean Nusselt number for the top surface, Nu_{top}, and the overall mean Nusselt numbers, Nu_{total}, both increase as the dimensionless plate width, W, decreases and the dimensionless gap between the plates, $VGap$, increases, the magnitude of this effect increasing as the Rayleigh number increases.

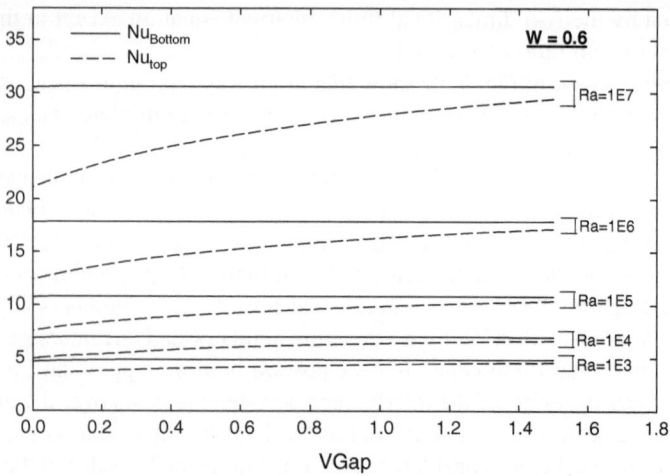

Fig. 7.21 Variation of mean Nusselt number for the *top* and *bottom* surfaces with the dimensionless gap between the plates for various Rayleigh numbers for $W = 0.6$

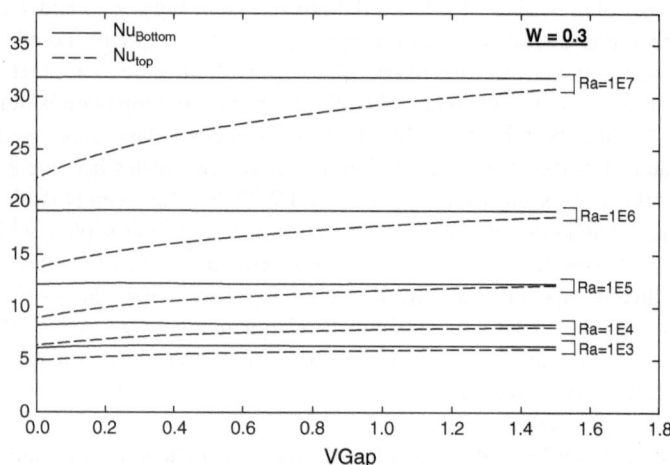

Fig. 7.22 Variation of mean Nusselt number for the *top* and *bottom* surfaces with the dimensionless gap between the plates for various Rayleigh numbers for $W = 0.3$

Because of the assumed form of the variation of Nu with Rayleigh number when edge effects are negligible, it has been assumed in attempting to correlate the results for the two vertical plate case that the Nusselt number for a plate of dimensionless plate width, W, is given by an equation of the form:

$$\frac{Nu}{Ra^{0.25}} = \text{constant} + \text{function } (Ra, W, VGap) \tag{7.7}$$

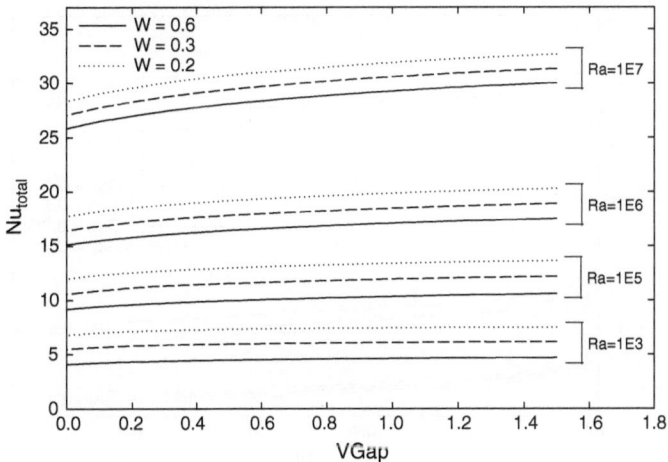

Fig. 7.23 Variation of average mean Nusselt number for the *top* and *bottom* surfaces with the dimensionless gap between the plates for various Rayleigh numbers and dimensionless plate width W

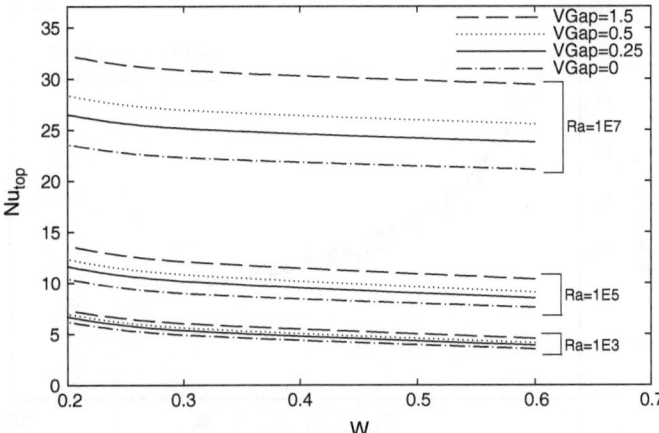

Fig. 7.24 Variation of mean Nusselt number for *top* surface with dimensionless plate width for various dimensionless gap *VGap* and Rayleigh number

Using an equation of this form, it has been found that the present numerical results for the case of two vertically separated plates with a dimensionless separation gap, *VGap*, between 0.1 and 1.5 can be approximately described by:

$$\frac{Nu_{\text{total}}}{Ra^{0.25}} = 0.45 + \frac{1.3}{\left[W \, Ra^{\left(0.3 + \frac{0.005}{VGap}\right)} \right]^{0.82}} \tag{7.8}$$

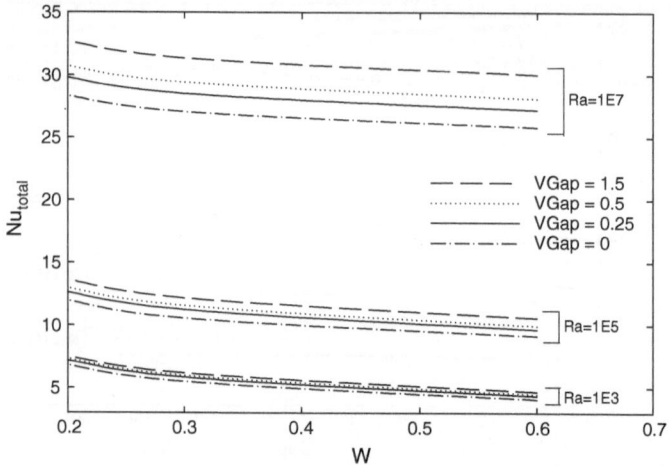

Fig. 7.25 Variation of the total mean Nusselt number for both *top* and *bottom* surfaces with dimensionless plate width for various dimensionless gaps *VGap* and Rayleigh number

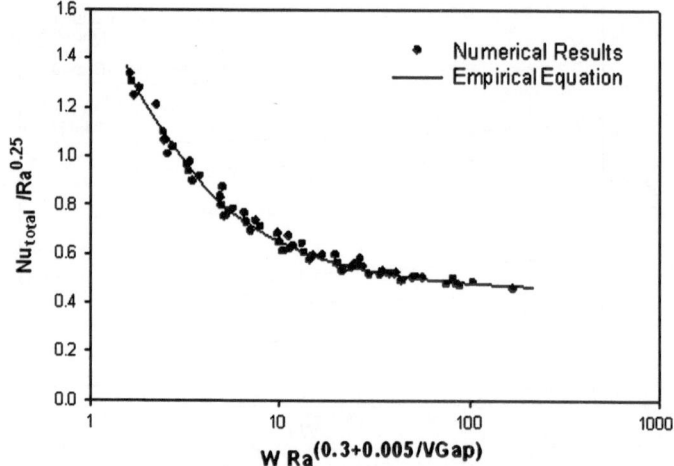

Fig. 7.26 Comparison of correlation equation for two narrow plates separated by a gap with numerical results

A comparison of the results given by this equation and most of the numerical results is shown in Fig. 7.26, the assumed correlation equation describing 80 % of the computed results to an accuracy of better than 10 %.

7.4 Conclusions

For the case of two horizontally adjacent plates, the heat transfer rate from each plate has been shown to be given by:

$$\frac{Nu}{Ra^{0.25}} - \frac{Nu_0}{Ra^{0.25}} = \frac{0.8}{\left[\left(\frac{W^3}{HGap}\right) Ra^{\left(0.5 + \frac{HGap}{2}\right)}\right]^{0.39}}$$

while for the case of two vertically adjacent plates the mean heat transfer rate from the two plates is given by:

$$\frac{Nu_{\text{total}}}{Ra^{0.25}} = 0.45 + \frac{1.3}{\left[W \, Ra^{\left(0.3 + \frac{0.005}{VGap}\right)}\right]^{0.82}}$$

References

1. Kalendar AY, Oosthuizen PH (2010) Natural convective heat transfer from two adjacent isothermal narrow vertical and inclined flat plates. JP J Heat Mass Transfer 4(1):61–80
2. Kalendar AY, Oosthuizen PH (2009) Natural convective heat transfer from two vertically spaced narrow isothermal flat plates. In: Proceedings of 17th Annual Conference CFD Society of Canada
3. Kalendar AY, Oosthuizen PH, Kalandar B (2009) A numerical study of natural convective heat transfer from two adjacent narrow isothermal inclined flat plates. Proceedings of ASME 2009 Heat Transf Summer Conference, CA. Paper HT2009-88091. doi: http://dx.doi.org/10.1115/HT2009-88091
4. Churchill SW, Chu HHS (1975) Correlating equations for laminar and turbulent free convection from a vertical plate. Int J Heat Mass Transfer 18(11):1323–1329